ANIMAL SCIENCE, ISSUES AND PROFESSIONS

CATS

BIOLOGY, BEHAVIOR AND HEALTH DISORDERS

ANIMAL SCIENCE, ISSUES AND PROFESSIONS

Additional books in this series can be found on Nova's website under the Series tab.

Additional E-books in this series can be found on Nova's website under the E-books tab.

ANIMAL SCIENCE, ISSUES AND PROFESSIONS

CATS

BIOLOGY, BEHAVIOR AND HEALTH DISORDERS

TANIYAH ZAKREVSKY
AND
VADIK LACHINOV
EDITORS

Nova Science Publishers, Inc.
New York

For permission to use material from this book please contact us:
Telephone 631-231-7269; Fax 631-231-8175
Web Site: http://www.novapublishers.com

Library of Congress Cataloging-in-Publication Data

Cats : biology, behavior and health disorders / editors, Taniyah Zakrevsky and Vadik Lachinov.
p. cm.
Includes bibliographical references and index.
ISBN 978-1-62081-043-9 (hardcover)
1. Cats--Diseases. 2. Cats--Physiology. 3. Cats--Behavior. I. Zakrevsky, Taniyah. II. Lachinov, Vadik.
SF985.C37 2011
636.8--dc23
2012002178

Published by Nova Science Publishers, Inc. † New York

CONTENTS

PREFACE

This book presents topical research in the study of the biology, behavior and health disorders of cats. Topics discussed include the diagnosis, treatment and prevention of parasitic infections in cats; development of the cat sympathetic ganglia in postnatal ontogenesis; feline perioperative analgesia; streptococcal disease as an emerging pathogen of cats; feline endometrial adnenocarcinoma; and nonspatial visual recognition memory in cats.

Chapter 1 - Cats as pets as stray are commonly affected by parasites, especially gastrointestinal, respiratory and skin parasites. Infections are usually associated with disease in kittens or in individuals with immunological disorders. In that sense, cats may act as a assimptomatic carrier playing a role in the maintenance of the parasite in the environment and increasing the risk to transmission to other cats or to human beings since some agents are considered zoonosis.

A revision of the most common parasites affecting cats would be presented:

Intestinal parasites would include protozoa such as *Giardia* sp., *Isospora* sp., and emerging pathogens such as *Tritrichomonas foetus* and *Cryptosporidium* sp. *Tapeworms*: Mesocestoides, Dipylidium sp. and Joyeuxiella sp, Taenidae (*Taenia taeniformis*, *Taenia pisiformis, Echinococcus multilocularis*). *Nematodes Ollulanus tricuspis. Roundworms*: *Toxocara cati*, *Toxascaris leonina, Hookworms*: *Ancylostoma tubaeforme*, *Uncinaria stenocephala*. With respect to respiratory parasites, *Eucoleus aerophilus*, *Dirofilaria* as an emerging pathogen in cats and some metastrongyloid nematodes have been reported from domestic cats, affecting respiratory tract. *Aelurostrongylus abstrusus* is rather common, nevertheless, others such as *Troglostrongylus* sp. and *Oslerus rostratus* have sporadically been reported.

External parasites would include ticks and mites (mange mites and *Cheyletiella*), fleas and flea-bite dermatitis and the chewing mite *Fellicola subrostrata*.

A good comprehension of the life cycle would be necessary to understand transmission, epidemiology and clinical presentation of the disease. Diagnosis of parasitic disease would be presented with special emphasis to the laboratorial diagnosis techniques and results. Control strategies would also be discussed including treatment, management and prevention.

Chapter 2 - In cats, sympathetic ganglia divide into para- and prevertebral. The largest paravertebral ganglia are the superior cervical and stellate. The superior cervical ganglion contains neurons that supply sympathetic innervation to the face, the stellate ganglion innervatethe organs of neck and thoracic cavity. Prevertebral ganglia composed of postganglionic sympathetic principal neurons that supply abdominal and pelvic viscera. Anatomical structure of the cat sympathetic ganglia is already formed to the moment of birth.

Nevertheless, elements of the cat stellate ganglion are not mature to the moment of birth. Maturation accompanies with the increasing of the neurons size and the decreasing of their number in the ganglion, with the reorganization of the connections of neurons in the ganglion. In postnatal ontogenesis in kittens, most of connections of the SG neurons with the cells-targets of viscera are formed in the first 2-3 weeks of life.

Electrophysiological properties also changes: conduction velocity and amplitude of discharges in sympathetic nerves increases, their forms become more complicated. Rhythmical discharges in sympathetic nerves are observed from the moment of birth and correlate with cardiac (1.2-3 Hz) and respiratory frequency (0.15-0.7 Hz). Amplitude of discharges is minimal in newborn kittens and increases during ontogenesis. Relative spectral power appears to be maximal at cardiac and respiratory frequencies in animals in all age groups. Morphological and functional processes of maturation pass heterochroneously and finish in kittens to six month of life.

Chapter 3 - Perioperative pain control has been increasingly implemented nowadays in cats undergoing various surgical procedures. However, effective analgesia provided in cats undergoing surgery is complicated because of the difficulties in pain recognition, unfamiliarity with analgesic techniques and drugs, limited availability of analgesics such as opioids, and lack of published studies evaluating the effectiveness of these drugs in the feline patients. Ovariohysterectomy, castration, onychectomy, limb amputations, thoracic, oncologic and reconstructive surgeries are among surgical procedures that require acute or chronic pain control. Perioperative multimodal and preemptive analgesia may include opioids, non-steroidal anti-inflammatory drugs (NSAIDs), α_2 agonists, ketamine and regional blocks. They may all provide safe analgesia, taking into account the differences in metabolism between cats and other species. Apart from routine pharmaceutical treatment novel surgical techniques have been devised to achieve pain relief in cats. This article reviews the physiology and pathophysiology of pain in cats undergoing minor and major surgical procedures and discusses pain diagnosis and management strategies for pain control in the feline patient. The effect of surgical technique on feline pain control is also reviewed.

Chapter 4 - Streptococcal disease is a common occurrence in animals. Most streptococcal species are host specific causing sporadic expression of disease in the host. *Streptococcus equi* subspecies *zooepidemicus* (SEZ) is a mucosal commensal in a wide range of animals, most significantly in horses, but is not a recognized commensal of the dog or the cat. Since the early 1980's, there have been increasing reports of SEZ linked to outbreaks of severe fatal respiratory disease in dogs maintained in shelters, research facilities, and kennels. The bacterium has been denoted an emerging pathogen of dogs kept in intensively housed environments. In 2010, a similar clinical presentation was reported in cats maintained in shelters. Similar to intensively housed dogs, sheltered cats exhibited severe fatal pneumonia, but were also reported to contract rhinitis and in some cases meningitis. SEZ is now considered an emerging pathogen of cats. In this chapter the authors discuss SEZ and expression of disease in the cat.

Chapter 5 - Endometrial adenocarcinomas are considered to be rare in domestic animals, in particular comparing to non-epithelial tumours of the uterus such as leiomyoma, in part because they are likely underdiagnosed. In cats, though uterus was the most common site for genital tract tumours, endometrial adenocarcinomas were found to be rare in all available references.

In the study presented herein, forty feline primary endometrial adenocarcinomas, identified by a minimum of three pathologists on conventional haematoxylin and eosin-stained sections, obtained from the archives of four different laboratories, for a period of 13 years were used in a retrospective study on the subject.

In previous studies on feline endometrial adenocarcinomas (FEA) it has been found an increase incidence in purebred animals, which has been associated with a longer reproductive activity in those animals in comparison with domestic cats. However, this should be read with caution, as population ratios between purebred and domestic mongrel animals may change with the geographic location. Also the cultural adoption of gonadectomy as a contraceptive measure may interfere with the regional prevalence of the disease. In fact, in Portugal the number of intact domestic cats is higher than the purebreds, due to the tendency for late spaying age (most of them due to uterine or mammary diseases), which could influence the incidence of FEA in the authors country.

Some of the specimens used in this study were obtained at ovariohysterectomy following a clinical diagnosis of pyometra. Unless metastasic disease exists, to which illness is often associated, most frequent signs for FEA are vague and unspecific, making difficult the early diagnosis of endometrial carcinoma.

The histological evaluation shows the existence of different tumour phenotypes, where clear cell are often found. The presence of nuclear atypia and the Ki67 expression seem not to be related to the malignancy of the tumour. Furthermore, one report refers the presence of oestrogen receptors, which could explain the low levels of metastization or invasiveness proposed for FEA, or the clinical outcome for the tumour. For most situations of primary FEA, in particular in early stages of the disease, myometrial invasion is limited and often absent. Also vascular invasion is seldom observed even in tumours with moderate nuclear atypia. However, limited information exists on specimens from early FEA stages, which impair the presentation of a reliable prognosis for that animal. The major challenge seems to be the identification of valuable prognostic markers to achieve a definitive prognosis for animal life.

Feline endometrial adenocarcinomas are possibly more common than the authors might presume and the clinical impact of such tumours may be increasing with the increase in the ability to establish an early diagnosis allied to an extended duration of life in cats. A recent report defends that FEA are more common than smooth muscle tumours of the uterus. Furthermore, as it frequently co-exists with pyometra or mucometra, it is also possible that a large number of cases fail to reach the pathological evaluation and hence remain undiagnosed.

In this chapter the authors propose to discuss the clinical and morphological data of all the forty feline primary endometrial adenocarcinomas, supported by an extended review of the literature

Chapter 6 - As it is well known in experimental psychology the evaluation of memories and their characteristics in people are produced by two basic methods: by the method of "free recall" and by the method of "recognition". As far as the authors know, for evaluating the memory in animals thus far effectively is used only the second of those methods. Since this part is dedicated particularly to memory in subhuman animals, speech subsequently will concern in essence precisely only the recognition memory, with small exceptions. At the same time the authors were deeply impressed by the success obtained by neuropsychologists in the USA in the study of recognition memory in animals (mainly in monkeys and rats), especially by the work of M. Mishkin's team at Bethesda laboratory of Neuropsychology.

Chapter 7 - The European wildcat (*Felis s. silvestris*) is one of the last small wild felines in Europe. Contrary to popular belief, it is not larger or heavier than most domestic cats (*Felis s. catus* also named *Felis catus*). Its weight has often been exaggerated in older publications due to confusions between pounds and kilograms or the voluntary omission of light specimens in analyses. The average weight of the European wildcat is around 5 kg for males and 3.5 kg for females even if the European wildcat seems bulkier and more robust than most domestic cats. The head-body length varies between 52 and 65 cm for males and 48.5 and 57 cm for females, and tail length varies between 26 and 34.5 cm for males and 25 and 32 cm for females.

In: Cats
Editors: T. Zakrevsky and V. Lachinov

ISBN 978-1-62081-043-9

Chapter I

PARASITES IN CATS

Anna Ortuño and Joaquim Castellà
Parasitologia i Malalties Parasitàries, Department de Sanitat i Anatomia Animals, Facultat de Veterinària, Universitat Autònoma de Barcelona, Barcelona, Spain

ABSTRACT

Cats as pets as stray are commonly affected by parasites, especially gastrointestinal, respiratory and skin parasites. Infections are usually associated with disease in kittens or in individuals with immunological disorders. In that sense, cats may act as a assimptomatic carrier playing a role in the maintenance of the parasite in the environment and increasing the risk to transmission to other cats or to human beings since some agents are considered zoonosis.

A revision of the most common parasites affecting cats would be presented:

Intestinal parasites would include protozoa such as *Giardia* sp., *Isospora* sp., and emerging pathogens such as *Tritrichomonas foetus* and *Cryptosporidium* sp. *Tapeworms*: Mesocestoides, Dipylidium sp. and Joyeuxiella sp, Taenidae (*Taenia taeniformis*, *Taenia pisiformis, Echinococcus multilocularis*). *Nematodes Ollulanus tricuspis*. *Roundworms*: *Toxocara cati*, *Toxascaris leonina, Hookworms*: *Ancylostoma tubaeforme*, *Uncinaria stenocephala*. With respect to respiratory parasites, *Eucoleus aerophilus*, *Dirofilaria* as an emerging pathogen in cats and some metastrongyloid nematodes have been reported from domestic cats, affecting respiratory tract. *Aelurostrongylus abstrusus* is rather common, nevertheless, others such as *Troglostrongylus* sp. and *Oslerus rostratus* have sporadically been reported.

External parasites would include ticks and mites (mange mites and *Cheyletiella*), fleas and flea-bite dermatitis and the chewing mite *Fellicola subrostrata*.

A good comprehension of the life cycle would be necessary to understand transmission, epidemiology and clinical presentation of the disease. Diagnosis of parasitic disease would be presented with special emphasis to the laboratorial diagnosis techniques and results. Control strategies would also be discussed including treatment, management and prevention.

Introduction

Cats can be infected with internal and external parasites. The incidence of parasitoses in cats is up to the level of awareness among owners, but, the particular behaviour of some cats that live outdoors and like roaming and hunting increase its risk to get infected. Therefore, feline populations such as catteries, feline colonies or breeding colonies provide suitable conditions for the occurrence and spread of parasitic infections, especially those transmitted by direct contact or fecal-oral route.

The use of broad-spectrum antihelmintics is quite common, but sometimes it is not an efficient strategy since parasite control is associated to something else than antihelmintic drug therapy. As Irwin suggested, there may be a perception that parasite problems in dogs and cats are generally well controlled as a result of advances in preventative drug therapy and the development of in-house tests for rapid diagnosis of infection and disease (Irwin, 2002). But, in the last decades, new parasitic diseases have emerged and re-emerged. This is the case of protozoan such as *Cryptosporidium spp.* or *Tritrichomonas foetus* or nematodes as *Dirofilaria immitis* that has been recently recognized in cats. Moreover, some parasites are easily transmitted and certain parasites can affect human beings. But not all cause significant clinical signs in cats. In some cases, cats are assimptomatic carriers but play a role in the transmission and maintenance of a parasite infection. In any case, it is important to detect them in order to establish suitable sanitary programs both for the animal itself as well as for those people who are likely to come into contact with it (children, owner, vet clinician, shelter staff, etc.). For that
purpose, a good understanding of the parasite biology, especially with respect to the life cycle and epidemiology is required for planning effective control strategies.

In this chapter, the most common parasitic infections are going to be review with especial emphasis to the diagnostic, treatment and prevention in order to give veterinary clinicians the tools to design efficient sanitary programs to parasite control.

Protozoa

Some of the most common protozoa causing diarrhea in cats are *Giardia* spp., *Isospora* sp., *Cryptosporidium* spp., and *Tritrichomonas foetus*. Furthermore, some protozoa affecting cats have a public health impact. This is the case of some *Giardia* assemblages or *Toxoplasma gondii* which is considered the causative agent of one of the most important zoonoses.

Giardia spp.

Giardia spp. is a Sarcomastigophora protozoa located at the intestinal tract of many animal species. The molecular characterisation of Giardia isolates has confirmed the existence of host specific species, such as the cat specific specie *Giardia felis* (assemblage F), and two

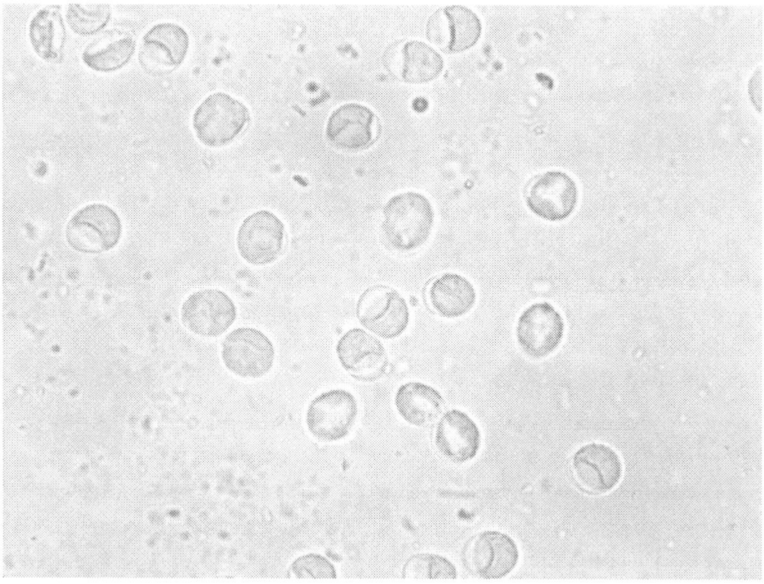

Figure 1. The cyst stage constitues the resistant stage, responsible for transmission, it is aproximately 12 µm long and 7µm wide.

species with broad host ranges and which are zoonotic; this is the case of *Giardia duodenalis* (assemblage A) and *Giardia enterica* (assemblage B) (Thompson and Smith, 2011)

The Parasite

Giardia has two stages in its life cycle. The teardrop-shaped trophozoite is the vegetative stage, with eight trailing flagella that is found in the intestinal tract. The ventral surface of the trophozoite bears a sucking disc. The cyst stage constitues the resistant stage, responsible for transmission, it is aproximately 12 µm long and 7µm wide (Figure 1). Cysts are highly resistant to environmental conditions and can survive several months in the environment in wet and cold conditions. Infection of Giardia is produced by ingestion of cysts contaminating food, water or environment.

Location in the Host

Intestinal tract.

Biological Cycle

Direct biological cycle. After cysts ingestion by the cat, excystation occurs in the duodenum releasing two trophozoites that adhere to enterocytes by means of the sucking disk, start feeding and establish an infection. The parasite alternates between attachment and free-swimming phases (Payne and Artzer, 2009). Trophozoites multiply repeatedly by binary fission. When the organism passes down through the intestinal tract, the surrounding internal environment changes and causes the transformation of the trophozoite to the cyst. The cysts are passed in the cat faeces and is considered to be the diagnostic stage of the parasite. It should be taken into account that cyst excretion from the host is intermittent. Trophozoites are not passed directly into the environment, but when diarrhea is severe or intestinal motility is extremely fast some trophozoites could be swept along. However, these trophozoites can not survive outside of the host and will not be infective.

Epidemiology

Fecal-oral transmission. Transmission occurs after ingestion of cysts present into the environment. Giardia cysts are highly resistant to environmental conditions and are shed in the feces intermittently and immediately infective.

Age and confinement can play an important role in Giardia infection, especially in a cattery setting (Papini et al., 2005). Moreover, factors related to the facility, such as cage material or floor material could also play a role in the maintenance of Giardia cysts, especially when those are constructed of porous material that retain humidity for a long time (Ortuño and Castellà, 2011).

Clinical Relevance

The typical clinical sign, when present, is diarrhea. It tends to be due to malabsorption and steatorrhea and strong odor is also present, occasionally with concomitant weight loss.

But, commonly, Giardia infection is assimptomatic in adult immunocompetent cats. In fact, it has been suggested that this organism is not always a primary pathogen since many infected cats are subclinical carriers.

Zoonoses

Despite *Giardia felis*, cats could be infected by G.duodenalis (assemblage A) that can infect human beings, too.

Cryptosporidium

Cryptosporidium spp. is an aplicomplexan coccidian. Several species have been described by means of morphological, biological and molecular data. Cats are usually infected by the host-specific *Cryptosporidium felis* although infections with other species have also been described.

The Parasite

C. felis is an apicomplexan intestinal coccidian. Oocysts passed in feces are sporulated containing four infective sporozoites. These oocysts are very small (4 -5 µm), so it can be difficult to detect microscopically. *Cryptosporidium* produces two types of oocysts. The majority are thickwalled, esporulated and passed in the feces, these are responsible to environmental contamination. Roughly 20%, are thin walled and release their sporozoites inside the intestine causing auto-infection.

Location in the Host

Cryptosporidium is located in the small intestine, at the tips of intestinal villi.

Biological Cycle

Cats are infected by ingestion of oocysts containing sporozoites. Sporozoites invade the microvillous brush border of the enterocytes forming a vacuole, intracellular but extracitoplasmatic. Within the vacuole, the parasite undergo asexual stage. Then go on to infect other cells and develops sexual stages. Oocysts are produced in 72h. While still within the host, oocysts undergoes a process of sporulation. Thus, host shed sporulated oocyst in feces that are immediately infective. These oocysts are shed into the environment but some of them, which has a thinwalled, excysts in the intestine, before to be shed and are responsible of the phenomenon of auto-infection.

Epidemiology

Cryptosporidium are transmitted via the fecal-oral route. Infection can mainly occur by ingestion of contaminated food or water or by grooming.

Oocysts are highly resistant to environmental conditions.

Clinical Relevance

Most *Cryptosporidium* spp. infected cats remain assimptomatic. Diarrhea is common in young animals, but preexisiting diseases such as feline leukemia or concomitant infections with other protozoans, especially *Giardia* spp or *Tritrichomonas foetus* may be associated with more significant illness.

Zoonoses

Infection in human beings are caused by *Cryptosporidium hominis* which is almost exclusively found in human or by *C. parvum* which is the etiological agent of zoonotic infections from cattle but there are some reports of infection caused by *C. felis*, too. (Meireles, 2010)

Isospora sp.

This intestinal coccidian affects dogs and cats. In cats, the two species recognized as a cause of intestinal coccidiosis are *Isospora felis* and *Isospora rivolta*. (Figure 2)

The Parasite

This coccidian shed unsporulated oocysts. The oocysts of *Isospora felis* (38-51μm x 27-39μm) are the largest of the coccidial oocysts affecting cats. The oocysts of *Isospora rivolta* measure 18-28μm x 16-23μm.

Oocysts sporulated in the environment, how long does it last to esporulate depends on the external conditions, especially temperature. Once sporulated, oocysts contain two sporocysts whith four sporozoites each one.

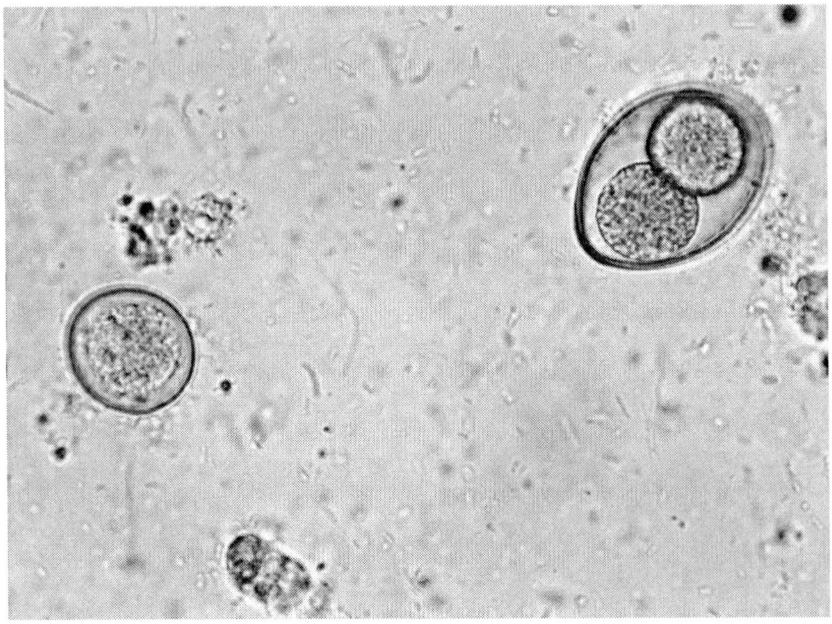

Figure 2. In cats, the two species recognized as a cause of intestinal coccidiosis are *Isospora felis* and *Isospora rivolta*.

Location in the Host

In the cat, the parasite is located in the small intestine but asexual stages are observed in extraintestinal tissues.

Biological Cycle

After ingestion of sporulated oocysts by cat, these excyst in the small intestine and sporozoites infect enterocytes of the distal portions of the villi in the ileum and rarely the duodenum and jejunum. Asexual stages are developed by endodyogeny and different types of meronts are produced. Later, sexual stages produce oocysts that are shed in cat feces unsporulated. Nevertheless, some sporozoites penetrate the intestinal wall and enter extraintestinal tissues where they form monozoic cysts in paratenic hosts for its entire life.

Epidemiology

Isospora is transmitted by fecal-oral route. There are two main routes of infection:

- Ingestion of sporulated oocyst by environmental contamination
- Predation of paratenic hosts

Clinical Relevance

Pathogenicity of the two *Isospora* species that affect cats is discussed controversely. Most infections are mild or subclinical, especially in adult cats. Clinical signs are most important in kittens or immunocompromised animals since immunological status plays an important role in the pathogenic process. In very young kittens, diarrhea or soft and mucoid feces are observed.

Zoonoses

No zoonotic capacity.

Tritrichomonas foetus

Tritrichomonas foetus is a protozoan parasite that recently has been recognized as an etiological agent of an emerging infectious diarrheal disease in cat, particularly in high-density populations.

The Parasite

Tritrichomonas foetus is a flagellated protozoa that has been considered for many years as a cause of venereal transmitted disease in bovines and recently it has been implicated as an intestinal pathogen in cats.

Location in the Host

Intestinal tract. It has also been detected in the feline uterus (Dahlgren et al., 2007).

Biological Cycle

Trichomonads do not have cyst stage. Trophozoites reproduce by binary fission within the intestine of the host. It is presumed the infection is caused by direct contact to the trophozoite since there is no resistant stage -cyst-. Nevertheless, a recent study observed that *T. foetus* trophozoites can survive and be potentially infectious in water, cat urine and dry cat food and canned cat food (Rossypal et al, Vet Parasitol 2011 *in press*)

Epidemiology

Affected cats are generally young (median age 8 months) and pure-breed cats.

High density is considered a risk factor.

Clinical Relevance

It is recognised as an important cause of chronic large-bowel diarrhea associated with blood, mucus, flatulence, tenesmus and anal irritation (Payne and Artzer, 2009). Concomitant infections with *Giardia* sp. are common (Bell et al., 2010).

Some *T. foetus* infected cats are asymptomatic.

Zoonoses

So far, it has not been considered to be zoonotic.

Toxoplasma gondii

This coccidian is one of the most important in terms of public health considerations. It can affect cats and a wide range of mammals and birds. The infection is worldwide distributed.

The Parasite

This coccidian is an Apicomplexan. Unsporulated oocysts measure 12x10μm. When sporulated, it contains two sporocysts with four esporozoites each one. Tachyzoites and bradyzoites constitued extraintestinal stages; both cells are banana shaped, tachyzoites are found in many cell types and bradyzoites are contained in cysts.

Location in the Host

As cats serve as both definitive and intermedite hosts, the intestinal stages of *Toxoplasma gondii* are located in the small intestine, mainly jejunum and ileum, and cats eliminate unsporulated oocysts in feces; the extra-intestinal stages are located in different tissues, especially the heart (Dubey, 1977).

Biological Cycle

Indirect biological cycle. As all coccidian, there is a sexual or enteroepithelial cycle and asexual or extraintestinal cycle. In the case of *Toxoplasma*, both cycles occur in the definitive host, the cat, while the asexual cycle occurs in non feline hosts –dog, sheep, people….etc.

Most cats become infected by ingesting intermediate host with tissue cyst containing bradizoites This phase penetrate the epithelial cells of intestinal wall and initiate the formation of schizonts. Merozoites released from schizonts transform into macro and microgametes. Microgamete fertilize a macrogamete, forming the zygot. An oocyst wall is formed around it. Unsporulated oocysts are passed in feces and sporulate outside the host, in the environment. When a hosts ingest a sporulated oocyst, sporozoites excyst in the intestinal lumen and become tachyzoites that multiply rapidly in any cell of the host; some of them begin to produce the latent tissue cyst satges containing bradizoites. Tissue cysts are formed in the spleen, lungs, liver, muscle or central nervous system and persist there for the entire life of the host.

Epidemiology

Although cat has been considered responsible of human toxoplasmosis, following infection cats shed oocysts for short periods of time (1-2 weeks), after which they become immune to reinfection. Furthermore, despite the high seroprevalence reported, the presence of *T. gondii* in feces is low (Dubey et al., 2009).

One of the most important risk factor is hunting and ingesting intermediate hosts, as rodents.

Clinical Relevance

Cats usually are not clinically ill.

Although *T. gondii* may act as an opportunistic pathogen when concomitant infections or immunosupression occurs, there is no evidence that infections caused by leukemia virus (FeLV) or immunodeficiency virus (FIV) modify the course of infection in cats (Dubey et al., 2009b)

When infection occurs transplacentally, clinical toxoplasmosis use to be severe. Kittens may be stillborn or die before weaning.

Postnatal infection causes lethargy, anorexia and dyspnea due to pneumonia. Other clinical signs include vomiting, diarrhea, icterus due to hepatitis, hyperesthesia on muscle palpation, fever, neurologic deficits etc.

Zoonosis

Cats are the source of infected oocysts. However, direct contact with cats is not thought to be a primary risk for human infection (Elmore et al., 2010). Human become infected through ingestion of oocyst-contaminated soil and water, tissue cysts in undercooked meat or congenitally.

Cat owners should be taken especial care when handling litter boxes, that should be changed daily and disinfected with boiling water. Pregnant women should avoid changing cat litter or use gloves when cleaning it or gardering or being in contact with soil.

PROTOZOA

Diagnosis

All cats with diarrhea should be assessed for enteric protozoal infection (Lappin, 2005). Diagnosis is based on detection of cysts, oocysts or trophozoites.

Some fecal filtration techniques could be available for some intestinal protozoa detection. Zinc sulfate centrifugation fecal flotation is indicated for detection of *Giardia* cysts and *Isospora* oocysts. Sugar centrifugation flotation would be suitable for detection of *Isospora* oocysts too, but distorts *Giardia* cysts because of its hipertonicity. In fact, Zinc sulfate centrifugation is considered the fecal flotation solution of choice for detection of *Giardia* cysts (www.capvet.org). But, it should be taken into account that *Giardia* shed cysts intermittently, so to increase the sensitivity of this technique, three fecal samples should be analyzed within 5 days.

With respect to *Cryptosporidium*, their small size and limited number of oocysts in feces of infected cats makes difficult diagnosis. So, acid-fast staining or fluorescein-labeled monoclonal antibodies in a Direct Immunofluorescence (IFD) of a thin smear is recommended. A commercially available IFD kit let us to diagnose *Cryptosporidium* and *Giardia*. Other
techniques, such as enzyme-linked immunosorbent assays (ELISA), are available to detect antigens of *Giardia* and *Cryptosporidium* in stool specimens.

A direct smear of diarrheic stool can be used to examine for trophozoites of *T. foetus*. But this technique has poor sensitivity. It' s recommended a in-clinic culture system that is commercially available to culture trichomonads from cat's feces.

With regard *T. gondii,* detection of oocysts in feline feces are best demostrated by centrifugation using Sheather sugar solution but are morphologically indistinguishable from other coccidia oocysts (*Hamondia* sp., *Besnoitia* sp). Furthermore, the oocyst shedding period is very short, so oocysts are rarely found on a routine fecal examination.

In general , diagnosis of clinical toxoplasmosis in cats can be based on the combination of serology and clinical parameters: high IgM titers or fourfold or greater IgG titers, exclusions of other causes of the clinical syndrome or beneficial clinical response to an anti-Toxoplasma drug (Dubey et al., 2009 (a))

Molecular techniques have been developed, showing high sensitivity but these techniques are not recommended to be used as a screening procedure and should be held in reserve for cases that are negative on other tests or for assessment of the zoonotic assemblage in the case of Giardia (Scorza and Tangtrongsup, 2010).

Treatment

Several treatment options have been reported and, although none has been shown to be universally effective, most cases can be successfully managed with drug therapy, supportive measures, and environmental control.

In *Giardia* infection, Metronidazole, Fenbendazole or a combination of Pyrantel, Febantel and Praziquantel are used most frequently in cats (Tangtrongsup and Scorza, 2010). Nevertheless, it is no possible to determine the optimal anti-Giardia drug since each case response different to the treatment. Current recommendations suggests a combination therapy with fenbendazole and metronidazole.

So far, there is no drug effective in eliminating *Cryptosporidium* from intestinal tract. However, infections can be self-limiting. Several drugs have been proved (Tylosin, Azithromycin, Paramomycin or Nitaxozanida) for weeks. Tylosin has positive effects but is not well-tolerated by most cats because of its unpleasant taste. Paramomycin decrease the number of oocysts shedding but it is unknown whether infection is eliminated.

Coccidiosis is generally self-limiting and cure spontaneously in healthy and immunocompetent cats. Nevertheless, administration of treatment may be necessary in kittens. Sulfadimethoxine or Trimethoprim-sulfonamides can be used. The efficacy of Toltrazuril and Diclazuril in the treatment of *Isospora* spp. infection in cats has been reported (Lloyd and Smith, 2001) showing a rapid clinical improvement and a fall in oocyst ouput.

Some *T. foetus* infected cats present coinfection with other protozoans. The only known effective treatment is Ronidazole that has led to parasitological cure and resolution of clinical disease, but should be used cautiously because can induce central nervous system toxicity (LeVine et al., 2011)

Concomitant infections should be taken into account in those cats with chronic diarrhea.

There is no approved treatment for clinical toxoplasmosis in cats. Clindamycin, Trimethoprim-sulfonamide or Toltrazuril have been use with varying results (Elmore et al., 2010).

Control and Management

Environmental contamination constitute one of the most important causes of no-response to drug treatment in protozoan infections. So, the most effective way to control and prevent protozoan infections is to establish control strategies in the environment:

1. Feces should be remove promptly from the environment. This is especially interesting in *Isospora* infection, since elimination of feces will avoid the sporulation

of *Isospora* oocysts or Toxoplasma gondii because oocyst takes more than one day after shed to get infective stage.

2. Steam cleaning to destroy oocyst and cysts. It is very important to avoid the use of water pressure hose because it may favour spreading cysts and ooocysts to other areas.
3. Disinfecting contaminated surfaces. The use of quaternary ammonium compound seem to be effective to Giardia and other protozoans. (Ortuño and Castellà, 2011). Disinfectants require enough contact time to be effective and organic debris could inactivate its effectivity.
4. Treatment and bathing of all in contact animals.
5. Control of paratenic hosts.
6. Feeding cats with commercial diets instead of raw meal.
7. Keeping cats indoors.

TAPEWORMS

Cestodes or tapeworms are platyhelminths consisting of a scolex, neck region and segments or proglottids. Their life cycles are indirect and generally occurs in the form of a predator-prey relationship. Cats, as predators, usually act as definitive hosts acquiring the adult form of the tapeworm by the ingestion of the larval metacestode stage contained in an intermediate host. The eggs contain an hexacanth embryo that is immediately infective.

Most common tapeworms affecting cats worldwide are *Dipylidium caninum*, *Mesocestoides* sp., *Taenia pisiformis* and *Echinococcus multilocularis*, a serious public health threat.

Mesocestoides spp.

The Parasite

The adult stage of *Mesocestoides* is about 30-70 cm. in length, the scolex has 4 muscular suckers without rostellum. In the gravid proglottids, eggs are contained in a parauterine organ.

Location in the Host

Cats, as definitive host, are infected by adult stages of *Mesocestoides* sp. that are located in the small intestine.

Biological Cycle

So far, the life cycle is unknown. But what is known is that it appears two intermediate hosts. The first intermediate host is assumed to be an arthropod, may be ants, that is infected by the cysticercoids stage. The second intermediate hosts are various vertebrates animals: amphibian, reptiles, birds and mammals (even the cat) that harbor the second larval stage, the tetrathyridium.

Cats acquire the infection by ingesting the metacestode stage (tetrathyridium) contained in the abdominal cavity of amphibian, reptile, birds or mammals. Tetrathyridium develops into a adult cestode in the cat's intestine but not all tetrathyridia ingested by cats develop to the adult stage, some migrate through to the intestinal wall to continue as the tetrathyridial stage in the cat (Bowman, 2002).

Epidemiology

Mesocestoides spp. is a cosmopolitan cestode. Prevalences reported ranged from<0.1% in Germany (Barutzki and Schaper, 2011) to 13.4% in Iran (Borji et al., 2011).

Since its biological cycle is not completely defined, the epidemiology of this parasite is not completely clarified but hunting activity of cats could be an important risk factor.

Clinical Relevance

No clinical signs are associated with the adult *Mesocestoides* in the small intestine of cats. In rare occasions, if cat harbors the tetrathyridial stage could develop a peritoneal cestodosis (Eleni et al., 2007).

Zoonoses

Human cases have been occasionally reported worldwide.

Dipylidium caninum

The Parasite

This tapeworm is 15 to 70cm. in length. The scolex is tiny with a four muscular suckers. It has an armed rostellum retractable into the scolex. Mature proglottids posesses bilateral genital pores. Gravid proglottids (10-12mm) contain egg capsules, each of which contain from 1 to 30 hexacanth embryos. These gravid proglottids have a cucumber seed appearance and passed singly in the feces. But they have the ability to move about the cat's perianal region, on the feces, on the bedding etc. When they arrive to the environment, they lose moisture resembling rice grains.

Location in the Host

The adult worms are located in the small intestine of its definitive host, the cat.

Biological Cycle

In this indirect life cycle are involved the definitve host, the cat, and the intermediate host: fleas (*Ctenocephalides felis*, *Ctenocephalides canis*, *Pulex irritans*) and the dog louse (*Trichodectes canis*). *D. caninum* is transmitted by ingestion of fleas or lice. Gravid proglottids are passed in the cat's feces in as little as 17 days after ingestion of infected fleas. The larval stages of the cat flea eat these proglottids. When the hexacanth embryo arrive to the intermediate host, develops into a metacestode: Cysticercoid that becomes infective when the adult flea has emerged from its pupal stage. The cat becomes infected by ingesting the flea during the grooming process.

Epidemiology
The main source of infection in cats is due to flea parasitation.

Clinical Relevance
Infections are well tolerated. Heavy infections can produce constipation and diarrhea

Zoonoses
It should be taken into account of the public health significance of *Dipylidium caninum.* When a flea containing infective cysticercoid stage are ingested by a human being, mainly children due to their close contact to cat pets.

Taenidae

Taenidae is a family of tapeworms that includes two important genera: G° *Echinococcus* and G° *Tenia*. Cats may be infected by *Taenia taeniformis* and *Echinococcus multilocularis*, although other taenidae could also infect cats but rarely.

Taenia taeniformis
Cats acquire infection of *T. taeniformis* by ingestion of strobilocerci contained in the tissues of rodents. Infected cats shed gravid proglottids in feces or the actively motile segments may exit through the anus on their own. Rodents become infected by ingesting eggs which then develop into strobilocerci in various tissues.

Echinococcus multilocularis
E. multilocularis affects dogs, wild canids such as foxes and cats; nevertheless, cats seem to be less suitable hosts than canids (Conboy, 2009). When a cat is infected by *E. multilocularis*, the adult stage of the tapeworm is located in the small intestine. Cats acquire infection from predating a rodent infected with alveolar hydatids, the metacestode stage.

Epidemiology
Roaming and hunting constituted the most important risk factors for Taenia spp. infection in cats.

Clinical Relevance
Clinical signs are not usually associated with infection.

Zoonoses
It has been reported human infection by *Taenia taeniformis*. Nevertheless, the most important zoonotic risk is caused by *Echinococcus multilocularis* since human beings act as intermediate hosts and can be infected by the metacestode stage, the alveolar multilocular hydatid that is highly invasive in the tissues

Tapeworms

Diagnosis

Diagnosis of tapeworms is based on detection and identification of gravid proglottids in cat's feces or by detection of eggs on fecal flotation analysis. It should be taken into account that eggs are only present in feces if gravid proglottids have been damaged or released after shedding. Therefore, fecal flotation has poor detection sensitivity.

Usually, the owner collect the segments present in feces and bring them to the clinic for identification. Proglottids can be examined for the presence of eggs which contained hexacanth embryo and identification is based on egg size and morphology. In the case of *Dipylidium caninum* it is recommended to flat the gravid proglottids in a slide with a drop physiologic saline and disperse the egg packets that are contained in the gravid proglottids; these egg packets are characteristics of *D. caninum* and usually could contain from 1 to 30 eggs. (Figure 3)

On the other hand, *Mesocestoides* spp. could be identify by flatting gravid proglottids between two glass slides and to detect the parauterine organ inside. Eggs are thin-walled and measure 30-40 μm in size.

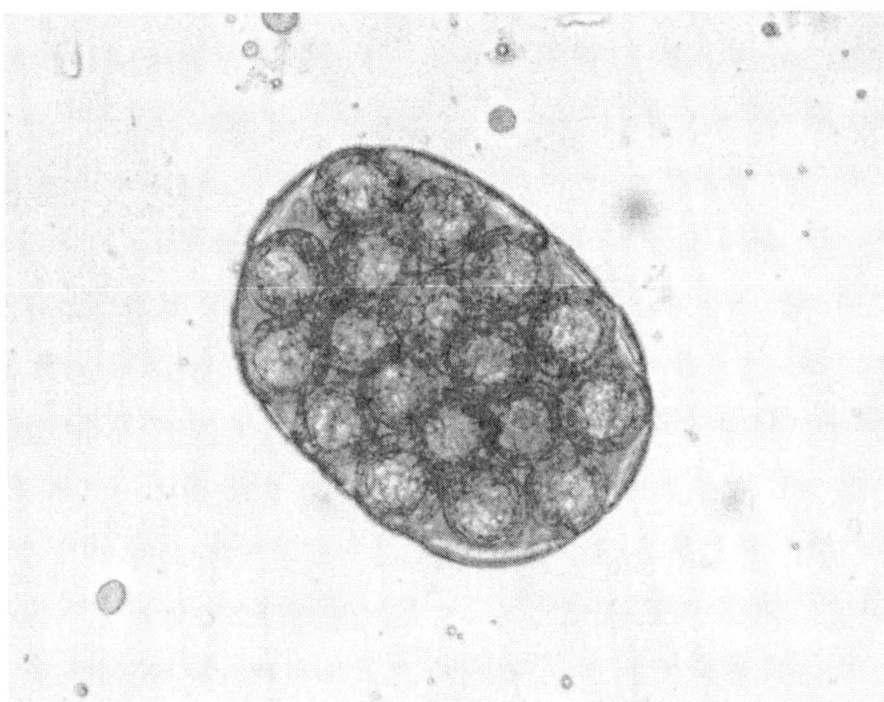

Figure 3. In the case of *Dipylidium caninum* it is recommended to flat the gravid proglottids in a slide with a drop physiologic saline and disperse the egg packets that are contained in the gravid proglottids; these egg packets are characteristics of *D. caninum* and usually could contain from 1 to 30 eggs.

In the case of Taenidae, eggs cannot be identify to a level of genus, since *Taenia* eggs cannot be diferenciate from those of *Echinococcus* sp. In that sense, detection of coproantigen and application of molecular biology techniques could be necessary in order to identify the parasite.

Treatment

Praziquantel or Epsiprantel are effective treatment for tapeworms in cats.

Treatment provokes releasing of a large number of infective eggs, so great care should be taken in the handling of these cats in the 72-hour post-treatment period.

Control and Management

Restriction of hunting and roaming activity would constituted one of the most important strategies to prevent tapeworm parasitation.

Application of a flea control program.

NEMATODA

Intestinal Worms

Toxocara cati

Toxocara cati is one of the most common helminth that affects cats, especially kittens. This ascaridida is a cosmopolitan parasite and prevalences ranged widely all over the world from 4,7% in Germany (Barutzki and Schaper, 2011) to 61.2% in Argentina (Sommerfelt et al., 2006).

The Parasite

The adult worms have three large lips and a glandular esophageal bulb. They have large cervical alae and are seen to have an arrow-head form. Adult worms have a length of up to 10 cm.

Eggs measure 65µm x 77µm and has a thick, pitted shell and a dimpled surface (Figure 4)

Location in the Host

Adults are found in the small intestine.

Biological Cycle

The adult worms are located in the small intestine. Females produce eggs that are passed in the feces of the cat containing a single cell, non-larvated. After a period of time in the

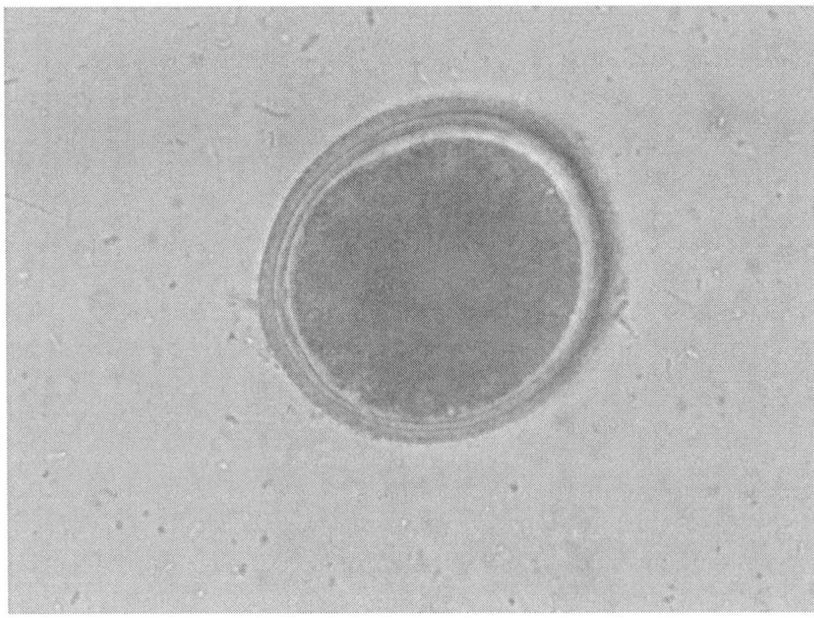

Figure 4. Eggs measure 65µm x 77µm and has a thick, pitted shell and a dimpled surface.

environment, eggs suffer two molts and produce the infective third-stage larvae. Following ingestion of larvated eggs or paratenic hosts, larvae arrive to stomach and intestine where molt to fourth-stage larva and, later, in the intestine lumen gets its preadult stage.

The migration pattern of *T. cati* differ qualitatively of those of *T. canis* since pre-natal infection with *T. cati* does not occur in the cat. The biology of *T. cati* leaves transmission of larvae via the milk as the only possible means of vertical infection. As it was observed by Coati et al. (2004), lactogenic transmission occurs after infection of the queen during late pregnancy. Following milk-borne infection, most of the larvae seem to undergo direct development in the intestine without tracheal migration (Coati et al., 2004).

Epidemiology

In general, neonatal infection via the mammary glands is an important route of infection in kittens and infected paratenic hosts represent an important source of infection for adult cats.

- Source of infection:
  - *Paratenic hosts:* Mice and other small host could play a role as a paratenic host of T. cati. Thus is an important source of infection in outdoors cats with a markedly predator character.
  - *Soil pollution: Toxocara cati* eggs are very resistant to environmental conditions and remain infective for years. Self-grooming constitues one way for ingestion eggs adhered to fur in cats.
  - *Lactogenic transmission* in kittens.

Opposite to *T. canis*, chronically infected queens wiuth *T. cati* harbor few somatic larvae (Lee et al., 2009); in fact, there is no evidence for the existence of arrested somatic larvae in the adult cat as an important host-finding strategy in the life cycle of *T. cati.*

Clinical Relevance

Clinical signs are associated to intestinal tract, especially in kittens highly infected: intermitent diarrhea, abdominal discomfort, vomitus (freqüently adult stages are present in the vomitus), poor coat, stunted growth.

Zoonoses

So far, *Toxocara canis* has been recognized as the most important cause of human toxocariosis. But, recently, *Toxocara cati* has also been implicated in human toxocariosis with symptoms associated to *Visceral Larva Migrans* and to *Ocular Larva Migrans* (Fisher, 2003) and it might have a greater role as zoonotic agent than estimated.

Toxascaris leonina

This ascarid affects cats but prevalences are, in general, lower than those detected in other ascarid infection such as *Toxocara cati* (Duarte et al., 2010; Barutzki and Schaper, 2011).

The Parasite

Female worms are around 6-15 cm. long and the males are around 5 cm. long. The cervical ale are longer and narrower than those of *T. cati.* The head of *T. leonina* resembles a spear while inthe case of *T. cati,* its head is similar to an arrow.

Eggs are ellipsoid, have a smoth shell and its dimensions are about 70µm x 80µm. They are pale and more traslucent than eggs of *Toxocara cati.*

Location in the Host

Small intestine

Biological Cycle

This ascarid tends to present a direct life cycle in cats; when the infective egg is ingested by a definitive host, it hatches in the stomach and the larva invades the mucosa of the small intestine where it developes and molts before returning to the lumen of the intestine to mature. If the egg is ingested by a paratenic host, the larva hatches the wall of the intestine where it remains for about one week before proceeding to other tissues where it encyst and remains arrested in the infective stage.

Vertical transmission of kittens is not described.

Epidemiology

- Sources of infection:

- *Eggs in soil.*_This is a very common source of infection. Eggs reaches its infective stage quickly, just one week, as compared to *Toxocara* sp. that need four weeks to developed an infective egg.
- *Paratenic hosts*

Clinical Relevance

Most infections are asymptomatic but, occasionally, enteritis have been observed (Epe, 2009).

Zoonosis

Toxascaris leonina is less well recognize as a zoonotic agent, as opposed to *Toxocara* sp.

Ancylostomatoidea

This group of parasites are commonly called hookworms because members of this family possess a dorsally flexed buccal cavity confering the characteristic hook posture on worm. Cats are affected by two genera: *Ancylostoma* and *Uncinaria*. The most commonly hookworms detected in cats are *Ancylostoma tubaeforme*, *Uncinaria stenocephala* and, in coastal areas of Africa and the Americas, A*ncylostoma braziliense* that also affects dogs. *Ancylostoma caninum* is regard as an uncommon parasite in cats.

The Parasite

The genus Ancylostoma is characterized by a large buccal capsule with marginal teeth, there being three pairs in *A. tubaeforme* and two pairs in *A. braziliense*. Specimens of *Uncinaria* possess cutting plates instead of teeth. It is a small worm, adults are 3-12 mm long.

Eggs of *Ancylostoma* and *Uncinaria* are indistinguisable from each other. Identification may be easy when mixed infection occurs since *Uncinaria* eggs are larger than *Ancylostoma*.

Location in Host

Small intestine.

Biological Cycle

Hookworm eggs are passed in the cat's feces and, under favorable environmental conditions such as damp, a rhabditiform larvae hatch from eggs. It molts twice an become a filariform larvae that constitues the infective stage, is present is feces or soil and infects cats orally or through skin penetration.

When skin penetration occurs, larvae migrate to the lungs, up the trachea and are swallowed, reaching the small intestine and mature into adults.

When a cat ingest L3, larvae arrive to the small intestine where they reach their adult stage.

Ancylostoma spp. are capable of being sequestered in the tissues of paratenic hosts, such as rodents, so infection by hunting is possible.

There is neither trasplacental nor transmammary transmission

Epidemiology

Risk factors for hookworm infection include climate, age or whether the cat had originated from a collectivity (cattery, feline colony, etc.) where it may have had a history of roaming and hunting and a greater exposure to contaminated environment (Palmer et al., 2007). With respect to the climate, warm temperatures provide optimal conditions for hookworm eggs hatching and larval development with the exception of *Uncinaria stenocephala* which requires lower temperatures for its development.

- Source of infection
  - Soil contamination. The main transmission route is through skin penetration.
  - Hunting. Paratenic hosts.

Clinical Rellevance

A. tubaeforme adult worms suck blood, causing blood loss and anemia, particularly important in kittens. However, *U. stenocephala* causes negligible blood loss, although sometimes can induce disease, especially in cases of severe parasitation.

Zoonoses

Hookworms can cause soil transmitted zoonosis, produce *Cutaneous Larva Migrans* (CLM) and less frequently eosinophilic pneumonitis, localized myositis, folliculitis, erythema multiforme or ophthalmological manifestations (Nevertheless, *Ancylostoma tubaeforme* larvae are considered minimal skin penetrators and appears not to cause CLM. On the other hand, the relationship between *U. stenocephala* and CLM remains unclear. (Bowman et al., 2010)

Intestinal Worms

Diagnosis

Roundworms and hookworms can be identified by faecal examination. Coprological techniques are suitable to detect ascarid or hookworm eggs. Identification of eggs is based on morphological characteristics, but in the case of hookworm species, identitification based on the morphology of the eggs recovered from the faeces is not appropriate as there is overlap in egg dimensions between most species of hookworm affecting dogs and cats. For that reason, it is required examination of the adult worms. However, it has been developed a highly species-specific and sensitive PCR-RFLP technique to detect and differentiate canine *Ancylostoma* spp. directly from eggs in faeces that identification. This technique allowed canine hookworm identification to be conducted rapidly, with ease and accuracy (Traub et al., 2004).

Treatment

Roundworms and hookworms should be treated with antihelmintics. Drugs such as milbemycin oxime, moxidectin, pyrantel and selamectin have been proved to be efficacious. (McTier, 2000; Schenker et al., 2007).

A single treatment of the queen with Emodepside/Praziquantel during late pregnancy can prevent the lactogenic transmission of *T. cati* to kittens and could therefore be taken as a useful tool in the strategic control of the parasite (Wolken et al., 2009).

Control

Regular coprological analysis and routinely deworming programmes should be applied in both kittens and adult cats; kittens should be dewormed at two, four, six and eight weeks of age as recommended by US Companion Animal Parasite Council (www. capcvet.org).

Related to the environmental contamination, daily removal of faeces from the floor costitues an effective preventive measure (Ortuño and Castellà, 2011). Furthermore, all surfaces should be physically clean, then sodium hipochlorite (1%) should be applied to strip off the outer protein coat of the ascarid eggs, but this measure is effective just when a preliminary proper cleaning has been done since the presence of residual organic matter will neutralize the disinfectant effect.

The role of paratenic hosts in the epidemiology of *T. cati* infection should always be taken into account in planning control measures.

Non-Intestinal Worms

Gastric Worms

Ollulanus tricuspis

This parasite has a worldwide distribution but there have only been sporadic reports. Colony-raised cats are more likely to be infected and, since it diagnosis is difficult it is possible that the occurrence of this parasite could be underestimated (Ortuño et al., 2011)

The Parasite

This trichostrongilid nematode inhabit the stomach of cats. The adult worms are very small, males are 0.7-0.9mm long and females are 0.8-1mm long.

Lyfe Cycle

This parasite has a direct life cycle and transmission from cat to cat appears to be direct through the consumption of infected vomitus (Bowman et al., 2002). Females are viviparous, releasing third stage larvae that are immediately infective, thus allowing the autogenous buildup of the worm population in the stomach. All stages present in the vomitus (larvae or adult) are capable to infect another cat. But, those specimen that enter the intestine do not appear to survive passage through the intestinal tract.

Location in the Host

Mucosa of the stomach.

Epidemiology

Free-roaming and colony-raised cats are more likely to be infected (Hargis *et al.*, 1983 b).

Source of Infection

Vomitus from other cats or from paratenic hosts.

Clinical Relevance

The pathogenicity of *O. tricuspis* is poorly understood. It has been considered clinically insignificant (Hargis *et al.*, 1983b); nevertheless, some authors describes the infection with an increased mucus production, anorexia and intermitent vomiting (Cecchi *et al.*, 2006). It has been sometimes associated with chronic gastritis, even with gastric adenocarcinoma in two cats (Dennis *et al.*, 2006).

Zoonoses

No zoonotic capacity.

Diagnosis

The antemortem diagnosis of infection is based on detection of different worm stages including larvae or adults in the vomitus, stomach contents or gastric mucosa. Only on very rare occasions diagnosis is made by direct fecal smears or fecal flotation methods (Hargis *et al.*, 1983a) since adults and larvae that enter the intestine are usually dead and destroyed (Bowman *et al.*, 2002).

Treatment

Tetramisole is reported to be an effective treatment (Haslinger, 1984)

Control

Specially in feline colonies and catterires, increased vomiting should be checked for the presence of *O. tricuspis* (Bowman, 2002).

Lungworms

Aelurostrongylus abstrusus

A. abstrusus is a parasite of the lung parenchyma in cats.

The Parasite

This parasite is a metastrongiloid whose biological cycle require a snail or a slug intermediate host. The copulatory bursa is of the basic strongylid pattern but has suffered varying degrees of reduction in the evolution of different families. Thus, family Angiostrongylidae, where belongs *Aelurostrongylus*, have a reduced bursa but the rays conform to the typical strongylid pattern and are well defined; the vulva is near the anus and the uterus is prodelphic.

Location in the Host

Alveoli and terminal bronquioli

Biological Cycle

Aelurostrongylus abstrusus female is oviparous and deposits eggs in "nests"in the lung parenchyma and first-stage larvae is developed. The first-stage larvae are carried up the tracheobronchial tree and swallowed, appearing in the cat's feces. These first-stage larvae enter any of a wide variety of snails and slugs. Two molts without cuticle shedding occur in the mollusc's foot tissues so that the infective larvae are enclosed in two sheaths. Cats may be infected experimentally by eating snails containing third-stage larvae, but the natural mode of infection is through predation paratenic hosts, such as rodents or birds, that normally eat snails or slugs. The third-satge larvae encyst in paratenic hosts's tissues and undergo no further development until they are ingested by a cat. Larvae appear in cat's feces five to six weeks after infection.

Epidemiology

- Sources of infection
  - Ingestion of paratenic hosts. *Aelurostrongylus* is present in cats with a marked predatory habit.

Clinical Rellevance

Most infected animals are free of clinical signs, but in moderate infections coughing and anorexia may be associated. Severe infections are manifested by cough, dyspnea and polypnea, all of which may be fatal.

Zoonoses

No zoonotic capacity.

Eucoleus aerophilus (Syn. Capillaria aerophila)

E. aerophilus is a trichuroid parasitic nematode that affects dogs, cats and wild carnivores. It is considered sporadic and most often sub-clinical.

The Parasite

Typically trichuroid, operculate eggs recognized by the morphometric and morphological characteristics, i.e. size (58–79 µm long x 29–40 µm wide), asymmetry of bipolar plugs and densely striated outershell.

Location in the Host

The adult lungworms live embedded in the epithelium of bronchioles, bronchi and trachea of the cat.

Biological Cycle

The biological cycle of *E. aerophilus* is not completely described. It is believed that cats are infected through ingestion of embrionated eggs. The females lay eggs that are coughed up, swallowed and released via faeces into the environment. Once eggs are deposited on the soil, they take about 30-45 days to reach the infective stage. Alternatively, development to the infective larval stage may occur in earthworms, as facultative intermediate hosts. Cats become infected by ingesting the embryonated eggs present in the environment or the earthworms. After ingestion, the larvae migrate to the lungs, where they evolve into adult stage and reach their sexual maturity after about 3–6 weeks post infection (Traversa et al., 2010)

Epidemiology

Although the epidemiological cycle of *E. aerophlius* is not still completely defined, hunting and paratenic hosts is suggested to be the most important source of infection.

Clinical Rellevance

Pulmonary capillariosis is typically characterized by a chronic bronchitis and infected animal may present a wide range of symptoms from minimal respiratory signs to inflammation, sneezing, wheezing, and chronic dry or moist and productive cough; (i.e. especially when bacterial complications intervene).

Zoonoses

Eucoleus infection has been esporadically reported in human beings.

LUNGWORMS

Diagnosis

Definitive diagnosis of *E. aerophilus* is by detection of eggs on fecal flotation or in bronchoalveolar lavage samples. The eggs are bipolar plugged and 58-79 x 29- 40 µm in size

With regard to *A. abstrusus*, diagnosis is by detection of L1 larvae in feces, bronchial mucus, or pleural fluid by Baermann examination. Faecal examination by Baermann technique is considered the gold standard for the diagnosis of *A. abstrusus*. It should be taken into account that sporadic shedding patterns can be present, so false-negative results can occur.

Treatment

Fenbendazole was shown to be effective in the treatment of aelurostrongilosis in cats. Furthermore, combination of Moxidectiba 1%-Imidacloprid 10% or Emodepside-Praziquantel have been proved to be safe abd effective in naturally infected cats.

With regard to *E. aerophilus* infection in cats, little information have been reported. Oral fenbendazole was effective in a dog infected and subcutaneous injection of abamectin was effective in a cat infected with *E. aerophilus* (Traversa et al., 2010)

Control

Control of cat hunting habits and paratenic hosts may constitue the most important control strategy. Moreover, regular coprological examination and deworming programmes should be apply in cats, especially outdoors cats.

Heartworms

Dirofilaria immitis

Feline heartworm infection, a vector-transmitted disease, is caused by *Dirofilaria immitis*. Although many cats are exposed to infection, not all develop adult stages since cats are considered accidental hosts. Thus, the prevalence of the heartworm infection in cats in a hyperendemic area is within the theoretically expected limits of 9-18% of the prevalence in dogs (Venco et al., 2011). In general, it has been recognised that cats living in areas where canine heartworm infection is endemic are at risk (Kramer and Genchi, 2002).

The Parasite

Adult worms are easily identified because of their location and their morphology.

Location in the Host

Adult worms are located in the pulmonary arteries and in the right cardiac chambers.

Biological Cycle

The biological cycle is indirect; mosquitoes of the family Culicidae act as vectors.

Female worms release microfilariae (L1) into the blood stream which are taken up by mosquitoes of the genera *Culex* sp., *Aedes* sp. and *Anopheles* sp., among others, during a blood meal. In the mosquito, L1 molt to the infective stage (L3). When the mosquito needs a subsequent blood meal, inoculates L3 to a new hosts.

Epidemiology

Maybe, the most important risk factor is the outdoor living cats since this disease is a vector-transmitted and it is necessary to be exposed to mosquitoes to be infected, although mosquitoes are not especially atracted by cats and it is considered that those cats living in hyperendemic areas for canine heartworm infection are in a greater risk.

Clinical Relevance

Cats have a natural resistance to infection. Most cats seem to tolerate *D. immitis* infection for extended periods. Some spontaneously self-cure as a result of natural death of parasites without demonstrating clinical signs, or they may suddenly show dramatic acute respiratory signs. Furthermore, sudden death in apparently healthy cats is a not rare event. Chronic symptoms include coughing, vomiting, diarrhea, and weight loss.

Zoonoses

D. immitis is zoonosis but parasite do not develop to adult stage in human beings since they are considered accidental hosts.

Diagnosis

Cats have a strong immune response againsts heartworms and this is reflected by low adult worms burden, low level of microfilaremia even amicrofilaraemic, prolonged prepatent period and short life span of adult stages (Venco et al., 2011). For all these reasons, diagnosis in cats is diffidult.

Detection of heartworm antigen test is commercially available. This technique (ELISA) have inherent limitations due to the great possibility of single-sex infections or a very low worm burden. Thus, in cats a negative result using antigen detection kit can not rule out the diagnosis of heartworm infection (Traversa et al., 2010).

Other diagnosis technques are detection of circulating antibodies againsts *D. immitis* and echocardiography.

Treatment

Studies have shown Melarsomine is toxic in cats at low doses. Ivermectine has shown some activity as an adulticide for use in cats. In any case, it should be taken into account that adulticide therapy is too risky due to the potentially severe complications associated with tromboembolism.

Treatment with tetracycline antibiotics prevent larval development, since it acts againsts Wolbachia, the endosymbiont bacteria of Dirofilaria that is involved in the development of imflamatory response and the polarization of host immune response to the parasite.

Control

Chemoprophylactic treatment is considered the safest measure for protecting cats in endemic/hyperendemic areas for canine HW infection.pot-on combination containing 10% imidacloprid/1% moxidectin (Venco et al., 2008) has been showed to be effective for the prevention of *Dirofilaria immitis* infection in cats.

References

Barutzki, D. and Schaper, R. Results of parasitological examinations of faecal samples from cats and dogs in Germany between 2003 and 2010. *Journal of Feline Medicine Surgery,* 2011; 13: 300-303.

Bell ET., Gowan RA., Lingard AE., McCoy RJ., Slapeta J. and Malik R. Naturally ocurring *Tritrichomnas foetus* infections in Australian cats: 38 cases. *Journal of Feline Medicine and Surgery,* 2010; 12:889-898.

Bowman, DD., Hendrix CM., Lindsay DS., Barr SC. *Feline Clinical Parasitology.* First Edition. Iowa State University Press: Blackwell Company Science; 2002.

Coati N., Schnieder T., Epe C. Vertical transmission of *Toxocara cati* Schrank 1788 (Anisakidae) in the cat. *Parasitology Research,* 2004; 92: 142-146.

Conboy G. Cestodes of dogs and cats in North America. *Veterinary Clinics Small Animals,* 2009; 39: 1075-1090.

Dahlgren SS, Gjerde B, Pettersen HY. First record of natural *Tritrichomonas foetus* infection of the feline uterus. *Journal of Small Animal Practice,* 2007; 48: 654-657.

Duarte, A., Castro, I., Pereira da Fonseca, IM., Almeida, V., Madeira de Carvalho LM., Meireles J., et al. Survey of infectious and parasitic diseases in stray cats at the Lisbon Metropolitan Area, Portugal. *Veterinary Parasitology,* 2010; 171 :163-6.

Dubey JP. Persistence of Toxoplasma gondii in the tissues of chronically infected cats. *Journal of Parasitology,* 1977; 63: 156-157.

Dubey JP., Lindsay DS. and Lappin M.(a). Toxoplasmosis and other intestinal coccidial infections in cats and dogs. *Veterinary Clinics of Small Animal,* 2009; 39: 1009-1034.

Dubey JP., Lappin MR., Kwok OCH et al.(b). Serorpevalence of *Toxoplasma gondii* and concurrent bartonella spp., feline immunodeficiency virus and feline leukemia infections in cats from grenada, west Indies. *Journal of Parasitology,* 2009; 95:1129-33.

Cecchi,R., Wills, SJ., Dean, R., Pearson, GR. Demonstration of O*llulanus tricuspis* in the stomach of domestic cats by biopsy. *Journalof Comparative Pathology,* 2006; 134: 374-377.

Dennis, MM., Bennet, N., Ehrhart, EJ. Gastric adenocarcinoma and chronic gastritis in two related persian cats. *Veterinary Pathology,* 2006; 43: 358-362.

Eleni, C., Scaramozzino, P., Busi, M., Ingrosso, S., D'Amelio, S. and De Liberato, C. Proliferative peritoneal and pleural cestodiasis in a cat caused by metacestodes of

Mesocestoides sp. *Anatomohistopathological findings and genetic identification.* Parasite-Journal de la Societé Française de parasitologie, 2007; 14: 71-76.

Epe C. Intestinal nematodes : biology and control. Veterinary Clinics Snorth *America Small Animal Practice,* 2009; 39: 1091-1107.

Elmore SA, Jones JL, Conrad PA, Patton S, Lindsay DS, Dubey JP. *Toxoplasma gondii*: epidemiology, feline clinical aspects, and prevention. *Trends in Parasitology,* 2010; 26: 190-196.

Fisher, M. *Toxocara cati:* an underestimated zoonotic agent. *Trends in Parasitology,* 2003: 19: 167-170.

Hargis AM., Haupt KH., Blanchard JL.(a) *Ollulanus tricuspis* found by fecal flotation in a cat with diarrhea. *Journal of American Veterinary Medical Association,* 1983; 182: 1122-1123.

Hargis, AM., Prieur, DJ., Blanchard, JL.(b) Prevalence, lesions, and differential diagnosis of *Ollulanus tricuspis* infection in cats. *Veterinary Pathology,* 1983; 20: 71-79.

Hasslinger MA. *Ollulanus tricuspis*, the stomach worm of the cat. Feline Practice, 1984; 14: 22-35.

Irwin P. Companion animal parasitology: a clinical perspective. *International Journal for Parasitology,* 2002; 32: 581-593.

Kramer L, Genchi C. Feline heartworm infection: serological survey of asymptomatic cats living in northern Italy. *Veterinary Parasitology,* 2002; 104:43-50.

Lappin MR. Enteric protozoal diseases. *Veterinary Clinics Small Animal,* 2005; 35: 81-88.

Lee, AC., Schantz, PM., Kazacos, KR., Montgomery SP. and Bowman DD. Epidemiologic and zoonotic aspects of ascarid infections in dogs and cats. *Trens in Parasitology,* 2010; 26: 155-161.

LeVine DN, Papich MG, Gookin JL, Davidson GS, Davis JL and Hayes, RB. Ronidazole pharmacokinetics after intravenous and oral immediate-release capsule administration in healthy cats. *Journal of Feline Medicine and Surgery,* 2011; 13: 244-250.

Lloyd S., and Smith J. Activity of Toltrazuril and diclazuril against Isospora species in kittens and puppies. *The Veterinary Record,* 2001; 21: 509-511.

McTier, TL., Shanks, DJ., Wren, JA., Six, RH., Bowman, DD, McCall, JW et al. Efficacy of selamectin against experimentally induced and naturally acquired infections of *Toxocara cati* and *Ancylostoma tubaeforme* in cats. *Veterinary Parasitology,* 2000; 91: 311-319.

Meireles MV. *Cryptosporidium* infection in Brazil: implications for veterinary medicine and public health. *Revista Brasileira Parasitologia Veterinaria,* 2010; 19:197-204.

Ortuño A. and Castellà J. Intestinal parasites in shelter dogs and risk factors associated with the facility and its management. *Israel Journal Veterinary Medicine,* 2011; 66: 103-107.

Ortuño A., Ferrer D., Gutierrez J., and Castellà J. First Report of *Ollulanus tricuspis*-infection in a cat in Spain. *Revista Ibero-Latinoamericana de Parasitología, in press.*

Palmer CS, Traub RJ, Robertson ID, Hobbs RP, Elliot A, While L et al. The veterinary and public health significance of hookworm in dogs and cats in Australia and the status of *A. ceylanicum. Veterinary Parasitology,* 2007; 145: 304-313.

Papini R., Gorini G., Spaziani A., Cardini G. Survey of giardiosis in shelter dog populations. *Veterinary Parasitology,* 2005; 124: 333-339.

Payne PA and Artzer M. The biology and control of *Giardia* spp. and *Tritrichomonas foetus*. *Veterinary Clinics of North-America,* 2009; 39: 993-1007.

Rosypal AC, Ripley A, Stockdale Walden HD, Blagburn BL, Grant DC, Lindsay DS. Survival of a feline isolate of *Tritrichomonas foetus* in water, cat urine, cat food and cat litter. *Veterinary Parasitology,* 2011; *in press.*

Schenker, R., Bowman, D., Epe, C., Cody, R., Seewald, W., Strehlau, G., and Junquera P. Efficacy of a milbemycin oxime–praziquantel combination product against adult and immature stages of *Toxocara cati* in cats and kittens after induced infection. *Veterinary Parasitology*, 2007; 145: 90-93.

Scorza V. and Tangtrongsup S. Update on the diagnosis and management of Cryptosporidium spp. Infections in dogs and cats. *Topics in Companion Animal Medicine,* 2010; 25: 163-169.

Sommerfelt, I.E., Cardillo, N., López, C., Ribicich, M., Gallo, C. and Franco A. Prevalence of Toxocara cati and other parasites in cats' faeces collected from the open spaces of public institutions: Buenos Aires, Argentina.*Veterinary Parasitology,* 2004;126 :249-55.

Tangtrongsup S. and Scorza V. Update on the diagnosis and management of Giardia sp. infections in dogs and cats. *Topics in Companion Animal Medicine,* 2010; 25: 155-161.

Traversa D., Di Cesare A. And Conboy G. Canine and feline cardiopulmonary parasitic nematodes in Europe: emerging and underestimated. *Parasites & Vectors, 2010*; doi: 10.1186/1756-3305-3-62.

Thompson RC. and Smith A. Zoonotic enteric protozoa. *Veterinary parasitology,* 2011; 182: 70-8

Venco L, Genchi M, Genchi C, Gatti D, Kramer L. Can heartworm prevalence in dogs be used as provisional data for assessing the prevalence of the infection in cats?. *Veterinary Parasitology,* 2011; 176:300-3.

VencoL., Mortarino, M., Carro, C., Genchi, M., Pampurini, F., Genchi C. Field efficacy and safety of a combination of moxidectin and imidacloprid for the prevention of feline heartworm (*Dirofilaria immitis*) infection. *Veterinary Parasitology,* 2008; 154: 67-7.

Wolken, S., Schaper R., Mencke N., Kraemer F. And Schnieder T. Treatment and Prevention of vertical transmission of *Toxocara cati* in cats with an Emodepside/Praziquantel spot-on formulation. *Parasitology Research,* 2009; doi 10.1007/s00436-009-1498-6.

Ectoparasites Affecting Cats

1. *Notoedres cati*

a. Introduction

Notoedres cati is a member of the family Sarcoptidae responsible for notoedric mange or feline scabies in cats. This mange is the veterinary importance and endemic in many parts of the world but uncomun in others.

The mite primarily attacks cats but may also infest foxes, dogs, rabbits and human which causes transient lesions (Foley, 1991; Bowman et al., 2001).

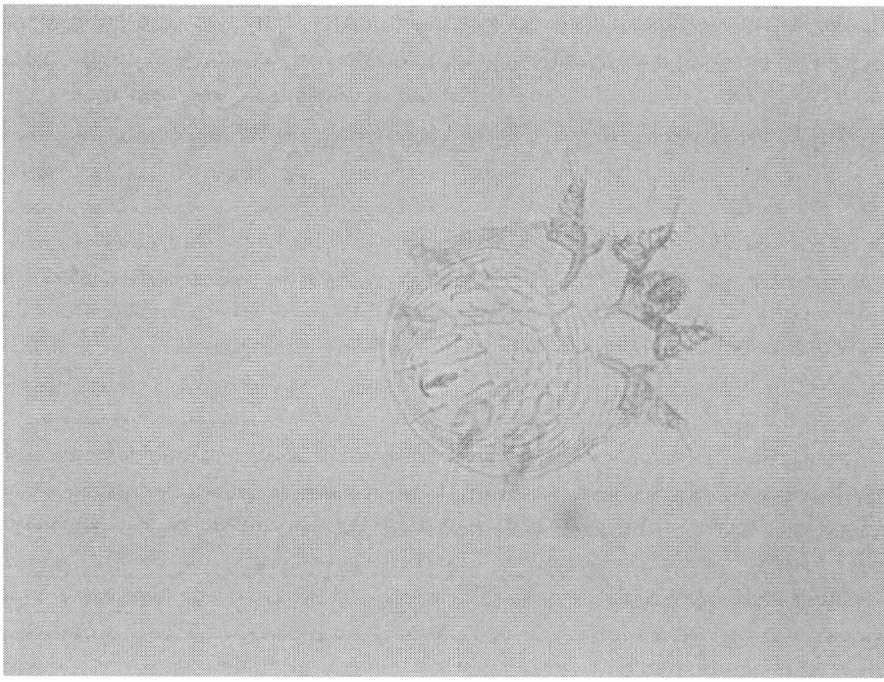

Figure 5. The mite resembles a little *Sarcoptes scabiei.*

b. The Parasite

The mite resembles a little *Sarcoptes scabiei,* but the concentric dorsal striations of the idiosoma, the dorsal anal opening and the absence of projecting dorsal scales make it possible their identification (Figure 5).

c. Location in the Host

As other mites of the family Sarcoptidae, *Notoedres cati* live in burrows in the epidermis.

d. Biological Cycle

The burrowing females laid eggs in skin tunnels. Larvae hatch and move to the surface on the skin and molt to the first nymphal stage. The nymph leaves the larval molting burrow and find another place on again undergoes ecdysis to become the second nymphal stage. In a third burrow the second nymphal stage molts to adult. The copulation occurs in these molting burrows when an stayed female is located by an male. The adults can appear as early as 12 days after the larvae hatches from the egg. The lifespan is thought to be about two months. The life cycle is completed in about 2 to 3 weeks. (Bowman et al., 2001; Arther, 2009)

e. Epidemiology

Mites of the genus *Notoedres* are found in cats, rabbits, mice, rats and bats. *Notoedres cati* is likely transmitted between domestic cats, another feline hosts, rabbits and some canids. The mites are highly infectious and transmission occurs primarily by animal-to-animal contact by larvae or nymphs (Wall and Shearer, 2001). There are several reports about household outbreaks of feline scabies in humans caracterized by a transient dermatitis (Bowman et al., 2001).

N. cati is an obligate parasite that survives off the host only a few days.

f. Clinical Relevance

Feline notoedric mange is a non-seasonal contagious ectoparasitic disease. Is noted as intensely pruritic, dry, crusted lesions that usually first appear on the ears, spread rapidly to rest of face and neck. The feet and perineum also can be affected. Skin appears thickened and alopecic, crusted and excoriated. Peripheral lymphadenomegaly is common. If untreated, lesions may spread and the affected animal can become severely debilitated, deaths can occur past several months (Paterson, 2008; Hnilica, 2011).

g. Diagnosis

Differential diagnosis should include autoimmune and allergic skin diseases and other ectoparasites as *Cheyletiella blackei, Otodectes cynotis, Fellicola subrostratus and Demodex spp.*

Diagnosis is based on clinical features and the identifications of mites and eggs in skin scrapings (Figure 6)

h. Treatment

Affected and all in-contact cats should be treated. Traditional treatments have included the use of lime sulfur or malathion dips. Topically applied selamectin has been used successfully. Alternative therapies include subcutaneous applications of ivermectin and doramectin (Ghubash 2006; Hnilica, 2011)

2. *Otodectes cynotis*

a. Introduction

Otodectes cynotis is a very common mange mite of cats, dogs and other carnivores, such as ferrets and foxes. The mite is found around the world, and it is often prevalent in animal shelters and breeding establishments (Bowman et al 2001).

b. The Parasite

Identification is very easy because *O. cynotis* is the only psoroptic mite tipically found in the ears of the cat. Female mites are 400 to 500 by 270 to 300 μm (Figure 7), whereas the males measure 320 to 400 by 210 to 300 μm (Figure 8)

c. Location in the Host

The parasites generally live deep in the ear canal but may be found in other areas of the cat body.

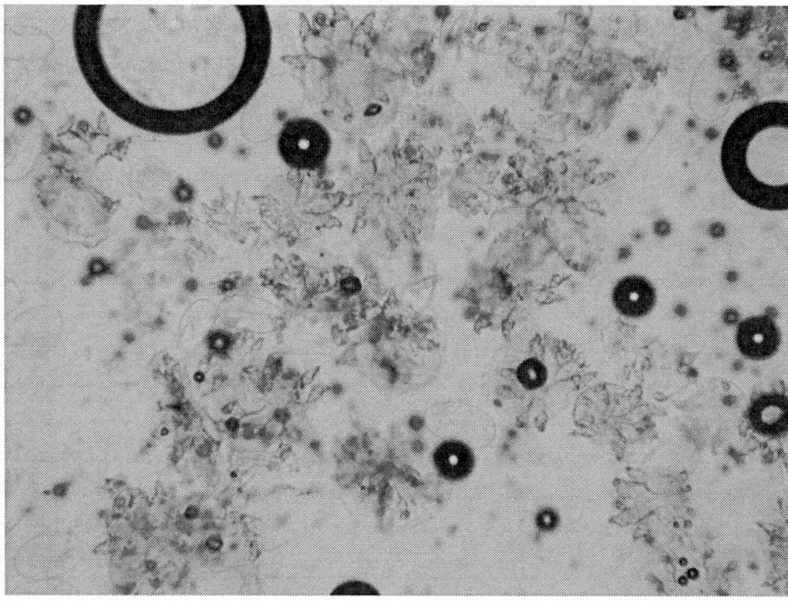

Figure 6. Diagnosis is based on clinical features and the identifications of mites and eggs in skin scrapings.

d. Biological Cycle

The life cycle of this permanent ectoparasite includes a larva, two nymphs stages and adults. The fertilized female laid and glued the eggs in the ear canal. The full life cycle took about 2-3 weeks (Wall and Shearer, 1997; Paterson, 2008; Arther, 2009)

e. Epidemiology

A considerable proportion of cats harbor subclinical infestations. Transmission between hosts is by direct contact. Probably, the infested females transferred the mites to kittens. *O. cynotis* is transmissible to other carnivorous pets (dogs and ferrets) but is uncommon the parasitism of humans (Bowman et al. 2001).

The mites seem to require an high relative humidity for survival and rapidly desiccate at typical room humidities. Under experimental conditions (35 C, 80% RH),can keep for months (Sweatman 1958, cit. Bowman et al, 2001).

f. Clinical Relevance

The mites cause irritation and local inflammatory reaction in the external ear canal by piercing the skin surface, feeding and migrating. The external ear canal in infested animals frequently has an accumulation of dark cerumen, dried blood and mite excretory products. Aural pruritus may cause the animal to rub, scratch and shake its head violently. It has been estimated that more than 80% of all cases of otitis externa in cats are caused by this mite. Occasionally ectopic infestation occurs where the mite causes localised dermatitis on the skin (neck, rump and tail) (Wall and Shearer, 2001; Arther, 2009).

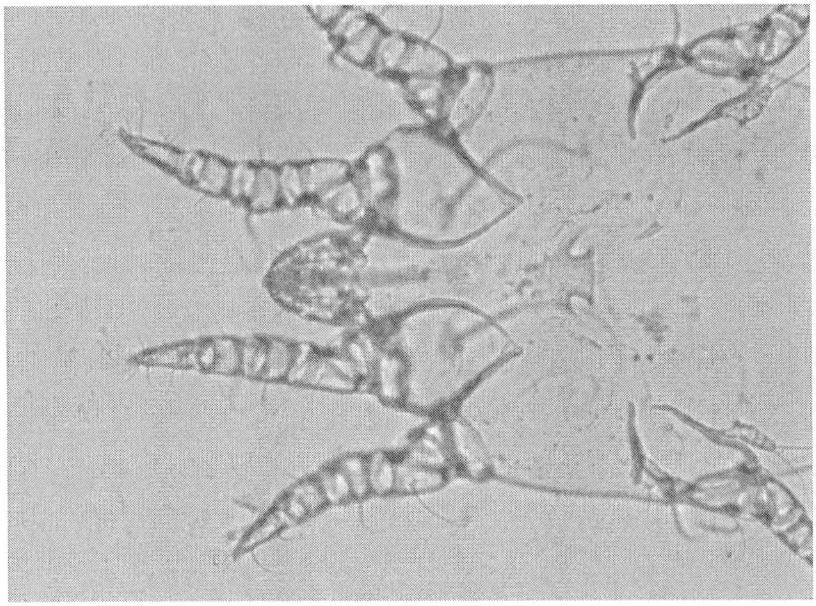

Figure 7. Female mites are 400 to 500 by 270 to 300 μm.

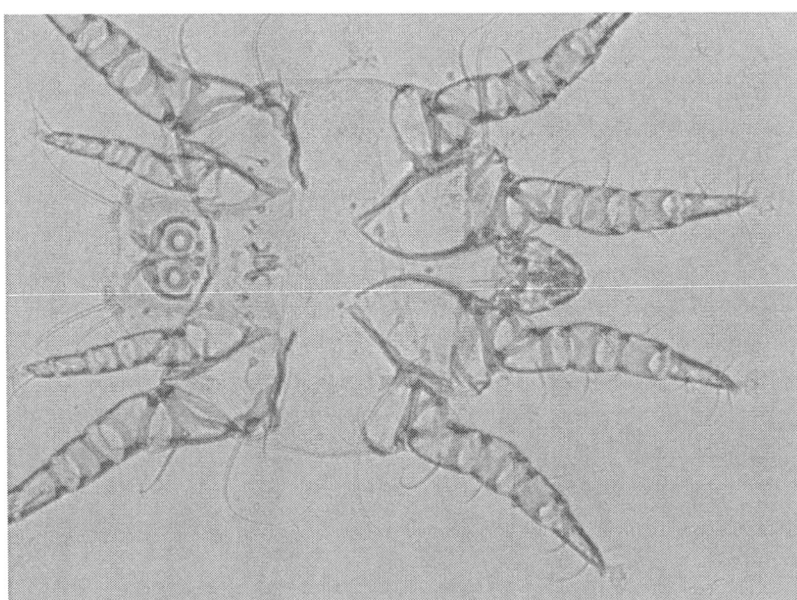

Figure 8. The males measure 320 to 400 by 210 to 300 μm.

g. Diagnosis

Mites may be indentified by direct examination of the ear canal using an otoscope or by swabbing the ear canal with a cotton applicator to remove the debris and observing this material under the microscope.

h. Treatment

Previous to acaricide treatment, the ear canals of affected animals should be cleaned to remove accumulated debris. An ceruminolytic agent for cleaning the zone may be convenient. This otoacariosis my be treated with Pyrethrin-containing compounds, rotenone-containing compounds and mineral-oil-based compound applied as otic suspensions. Ivermectin (0,001%) applied as an otic suspension controls mites and prevents the hatching of larvae from eggs. Efficacious systemic topically applied products are imidacloprid + moxidectin and selamectin (Ghubash, 2006; Hnilica 2011).

When an outbreak of otoacariosis is detected, all cats in the household should be treated and the equipments and bedding should be disinfected.

3. *Demodex* spp.

a. Introduction

The Demodicidae is a family of prostigmatid mites containing a single genus of veterinary interest, *Demodex*. Species of the genus *Demodex* are highly specie specific and frequently are found in animals and humans. The most part they are non-pathogenic and form a normal part of the skin fauna. Sometimes they can cause a clinical disease named demodicosis or demodectic mange.

b. The Parasite

Species of *Demodex* are minute mites with an elongated form (100-400 μm). The legs are stumpy and located at the front of the body (Figure 9). The body form in some species is elongated but in others is very short. Two species of Demodex can parasitize the cat, the long follicular mite *D. cati* and the short form *D. gatoi* (Wall and Shearer, 2001).

c. Location in the Host

D. cati is usually localised in hair follicles and sebaceous glands of skin, affecting the periocular skin, eyelids, head and neck. *D. gatoi* probably lives on the epidermal suface rather than in hair follicles (Desch and Stewart,1999)

d. Biological Cycle

It is assumed that *D. cati* female lay eggs within the epidermal cavity, and the eggs hatch to produce a six-legged larval stage. The larva then develops to a six-legged protonymph, and finally to an eight-legged nymph. Adult stages then follow. All life-cycle stages may be found together in a follicle (Wall and Shearer, 1997).

e. Epidemiology

Little is known about the epidemiology of feline demodicosis. *D. cati* is suspected that is difficult to transmit between cats unless there is prolonged contact. In nature such contact occurs only during suckling (Urquhart et al, 1987). The mode of transmission for *D. gatoi* is unknown but is probable that it may be contagious between adult cats.

f. Clinical Relevance

Demodicosis is a rare feline dermatosis, usually *Demodex* mites are commensals of normal feline skin. Demodectic mange due to *D. cati* is usually localized, affecting the periocular skin, eyelids, head and neck with erythema, scaling, crusts and alopecia. Generalised forms can be observed associated with underlying debilitating diseases. Demodicosis due to *D. gatoi* occurs in the absence of underlying diseases. The lesions affecting the head, neck and elbows can manifest with marked pruritus, alopecia, scaling, excoriation and crusts (Wall and Shearer, 2001, Paterson, 2008)

g. Diagnosis

Demodicosis is diagnosed on the basis of clinical features and identification of mites in deep and superficial skin scrapings. *D. gatoi* may be difficult to find.

h. Treatment

The localized forms by *D. cati* may resolve spontaneously o by topical aplications of amitraz solution. In the generalized forms, lime sulfur dips, sistemic doramectin and amitraz solution may be effectives. Treatments should be continued until lesions have resolved and the skin scrapings are negative for mites (Arther, 2009; Ghubash, 2006).

Any predisposing factor should be identified and corrected. The prognosis for localized forms is good, in the generalized demodicosis by *D. cati* is tied to the underlying cause. (Hnilica, 2011).

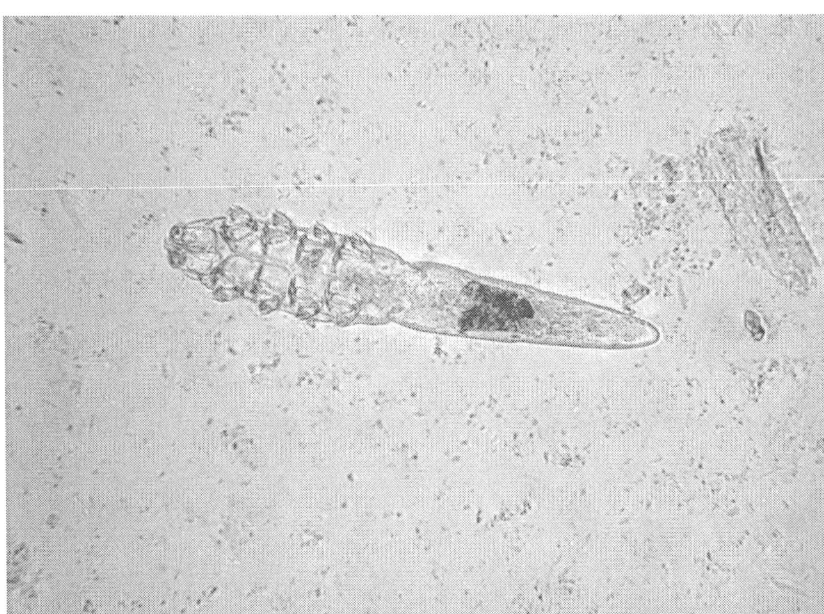

Figure 9. The legs are stumpy and located at the front of the body.

4. *Cheyletiella Blackei*

a. Introduction

Many species of the family Cheyletiellidae are free-living predatory, but some species of the genus *Cheyletiella* are ectoparasites. Three species of this genus are common in domestic animals: *C. yasguri* in dogs, *C. blackei* in cats and *C. parasitovorax* in rabbits (Urquhart et al. 1987).

Cheyletiellosis by *C. blackei* is a contagious infestation of the cat and particularly affects young animals. Cheyletiellosis has been referred to as "walking dandruff" (Wall and Shearer, 2001; Paterson 2008)

b. The Parasite

The adults are large mites (500 by 265 μm) visible to the naked eye. The palps are greatly enlarged and each ends in a prehensile claw (Figures 10,11). The palpi assist the mite in attaching to the host as it feeds on tissue fluids and skin surface debris. The legs terminate in "combs" instead claws or suckers.

c. Location in the Host

The mites live in the hair an fur and only visit the skin to feed. All stages of this mite are surface dwelling, they do not burrow but pierce the skin. Cats typically exhibit scaling on the head, neck and dorsum.

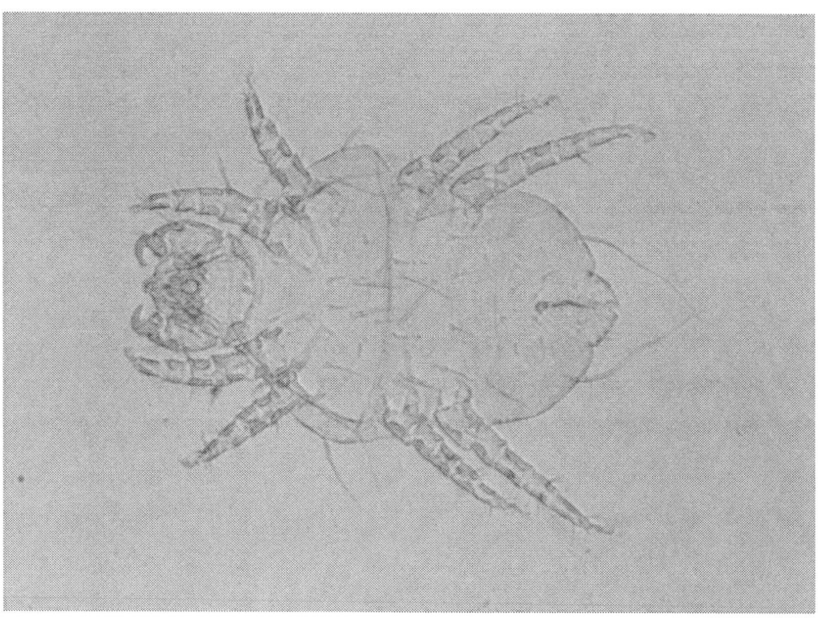

Figure 10. The palps are greatly enlarged and each ends in a prehensile claw.

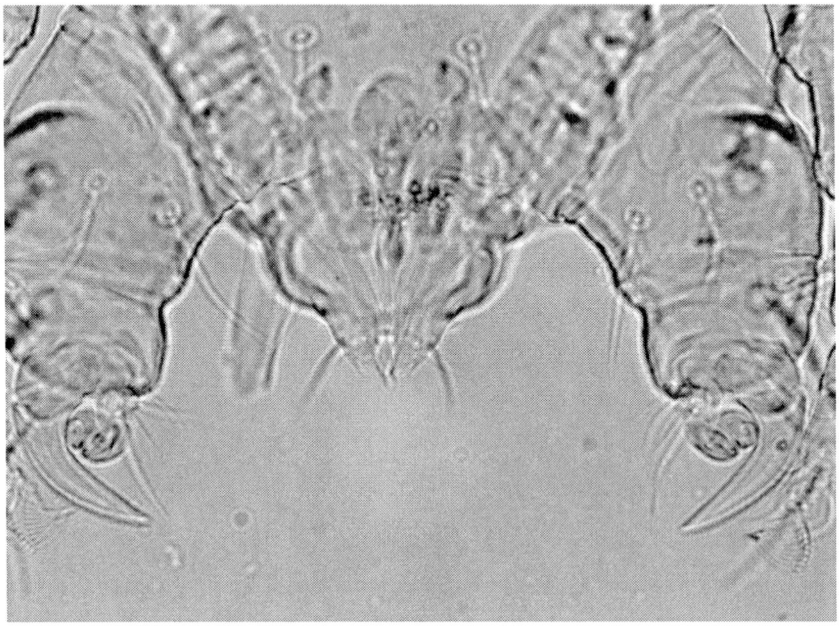

Figure 11. The palps are greatly enlarged and each ends in a prehensile claw.

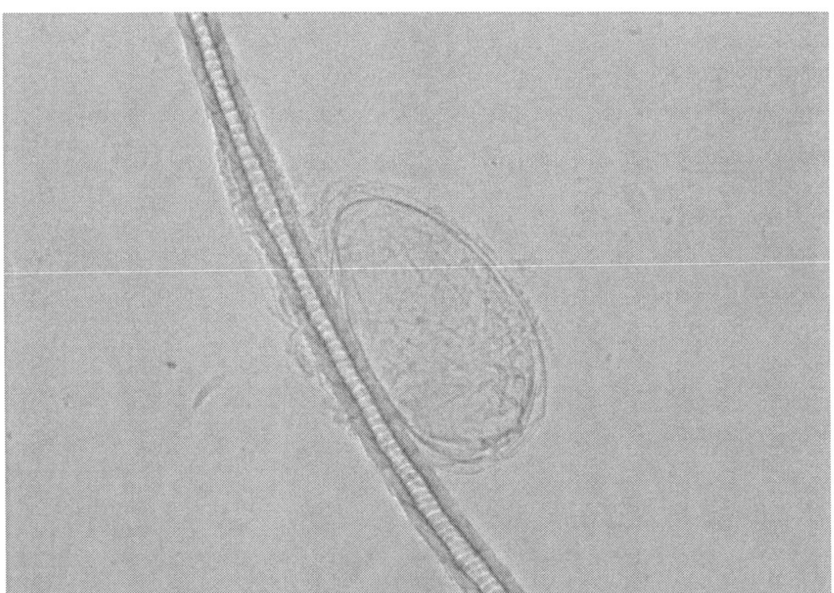

Figure 12. Eggs are attached to the hairs by the female mite.

d. Biological Cycle

Eggs are attached to the hairs by the female mite (Figure 12), a prelarva and larva develop within the eggshell. Fully developed nymphs emerge from the egg. There are two nymphal stages before the adult stage is reached. The life cycle is completed in about two weeks (Wall and Shearer, 2001; Arther 2009).

e. Epidemiology

This highly contagious, but low pathogenic mange, can spread rapidly in catteries. Adults can remain alive for up 1 month off their host without feeding.

The incidence is higher in kittens and long haired breeds. The mite is not strictly host specific and can affect rabbits, dogs and humans. *C. blackei* cannot complete its life cycle on humans (Bowman et al., 2001; Arther, 2009).

Cheyletiella infections seem to occur commonly in certain regions and may vary in frequency from year to year (Hnilica, 2011).

f. Clinical Relevance

The most common symptom is excessive scaling on the back, which gives the hair coat a powdery appearance. Pruritus is variable. Some times can evolve to a papulocrustous dermatitis affecting the dorsum. Asymptomatic carries can occur.(Paterson, 2008).

g. Diagnosis

Direct visualization of mites is possible by comb out dandruff onto dark paper, and observe for movement of mites in debris. The mites can be recovered by acetate tape impressions or superficial skin scrapings.

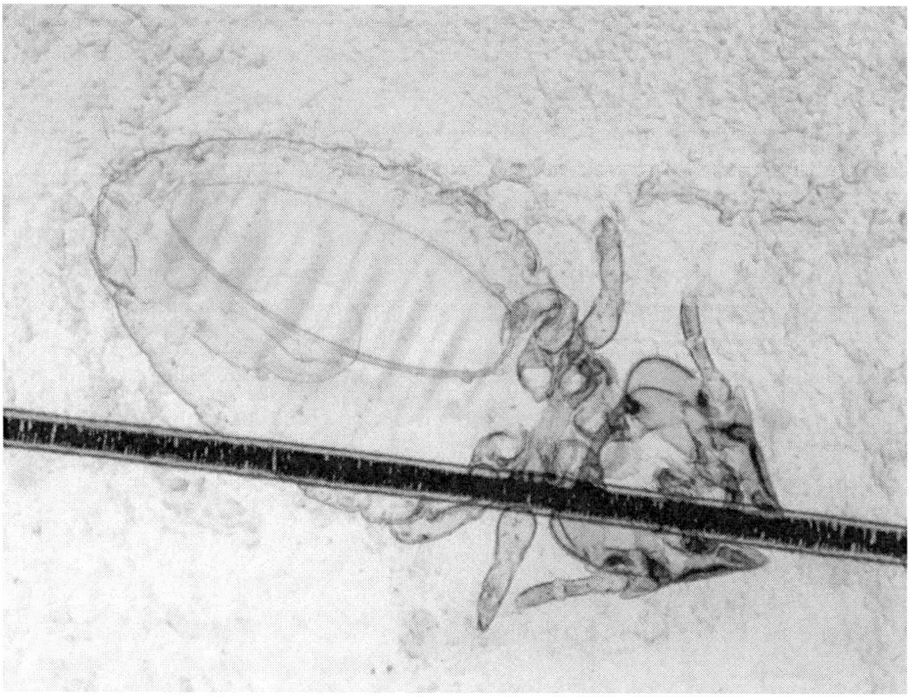

Figure 13. It is readily identified by the triangular shape of the head.

h. Treatment

Formulations containing pyrethrins (spray, shampoo or spoot-on), fipronil (spray or spoot-on) and selamectin (spoot-on) provided efficacy for the treatment of feline cheyletiellosis. The environment sholud be cleaned and treated with an insecticide (Ghubash, 2006; Paterson, 2008; Arther, 2009)

5. *Felicola Subrostratus*

a. Introduction

Feline pediculosis is an uncommon pruritic disease caused by the mallophaga louse *Felicola subrostratus.* The lice is found throughout the world.

b. The Parasite

The sole louse of any importance in cats is the biting species *F. subrostratus,* It is readily identified by the triangular shape of the head (Figure 13) and measure 1 to 1,5 mm (Arther, 2009)

c. Location in the Host

F. subrostratus live on the fur of the cat. The eggs are glued by the female to the hair.

d. Biological Cycle

The biology of *F. subrostratus* is very poorly known. The female glues her operculate eggs to the hair. From the egg hatch a nymphal lice that molt in several days. Adults may be present after 2 to 3 weeks. The longevity of adults is unknown (Bowman et al. 2001).

e. Epidemiology

The species seem highly host specific, but it is possible others felines can be infested. The humans and dogs are no affected by this louse.

Lice are more common in catteries where asymptomatic carriers may act as reservoirs. It is suspected that the infestation will be more common in the winter than in the summer (Bowman et al. 2001)

f. Clinical Relevance

Pediculosis is uncommon in cats, with highest incidence reported in neglected and underfed kittens.The clinical signs associated are pruritus, scaling, crusting, coat matting, excoriation and secondary alopecia (Pinchbeck and Hillier, 2006; Paterson 2008)

g. Diagnosis

Differentials include fleas, scabies, cheyletiellosis and hypersensitivity. The diagnostic includes direct visualization of lice and nits (operculate eggs glued to hair) and to take samples for microscopic examination.

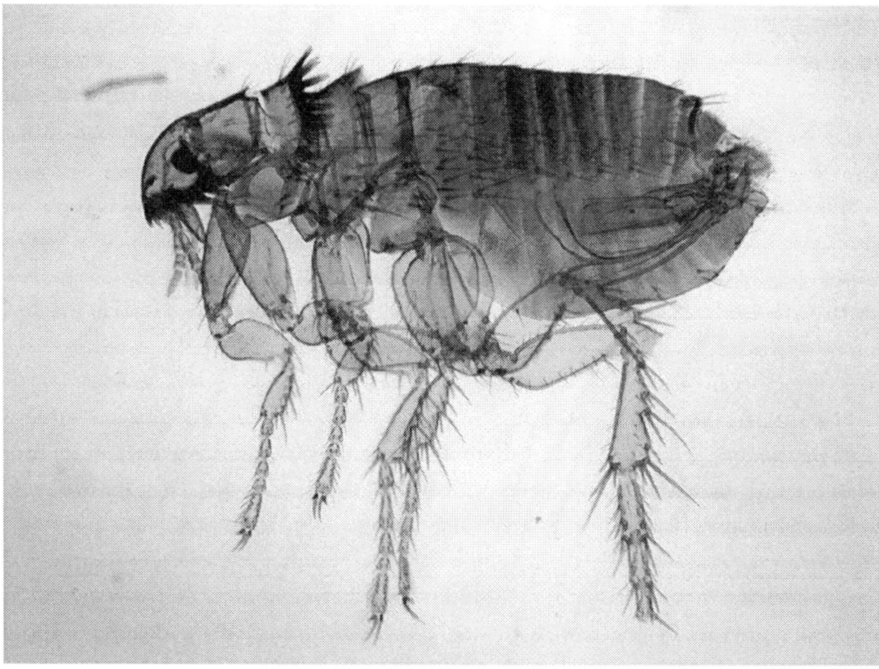

Figure 14. The cat flea has both genal and pronotal ctenidia.

Figure 15. In the female, the head is twice as long as high and pointed anteriorly.

h. Treatment

Subjects affected and all in-contact cats should be treated. The infested hairs with nits must be clipped away. Topic treatments with lime sulfur or pyrethrin can be used in cats. Alternative products include: Ivermectin, selamectin, doramectin and fipronil. Bedding, grooming tools, and environment should be cleaned. Insecticide flea collars may protect exposed animals (Pinchbeck and Hillier, 2006; Paterson, 2008; Hnilica, 2011).

6. *Ctenocephalides Felis*

a. Introduction

Fleas are small, wingless, blood-sucking insects. They are the most common cause of skin disease (flea allergy dermatitis) in dogs and cats worldwide. *C. felis felis* is an important intermediate host of the most frequently tapeworm of dogs and cats, *Dipylidium caninum.*

b. The Parasite

The major flea of importance to the cat are *Ctenocephalides felis felis* (cat flea). Occasionally, cats will also become infested with *Echidnophaga gallinacea* (poultry flea), *Spilopsyllus cuniculi* (rabbit flea), *C. canis* (dog flea) and *Pulex irritans* (human flea) (Paterson, 2008). An iconographic key to the differentation of fleas of veterinary importance is avaiable at Urquart et al. 1987.The cat flea has both genal and pronotal ctenidia (Figure 14). In the female, the head is twice as long as high and pointed anteriorly (Figure 15).

c. Biological Cycle

After locating an host, adult females mate and feed. Within 24-48 hours of the first blood meal females being to oviposit (30-50 eggs by day). The eggs are deposited on the host, but fall to the ground and tend to accumulate in areas where the host sleep or rest. The hatched larvae are free-living in all three larval instars (Figure 16). The mature third-stage larva spins a thin silk cocoon. Within the cocoon the larva pupates. When fully developed, adults emerge but may remain within the cocoon. Mechanical pressure, vibrations or heat, can induce emergence of the adults forms. Under domestic conditions, *C. f. felis* will complete the life cycle in 3-5 weeks. However, under adverse conditions this can be extended to 190 days (Dryden and Rust, 1994; Krämer and Mencke 2001)

d. Clinical Relevance

Flea bite dermatitis is one of the most common feline dermatoses. The clinical signs include: Symmetrical alopecia (by excessive grooming), pruritic miliary dermatitis (on the dorsum, tail base and the neck) and “Eosinophilic granuloma complex” (Paterson, 2008).

Non-flea-allergic cats may have no symptoms, or they may be anemic, have tapeworms (*D. caninum*), or develop mild skin irritation (Hnilica, 2011).

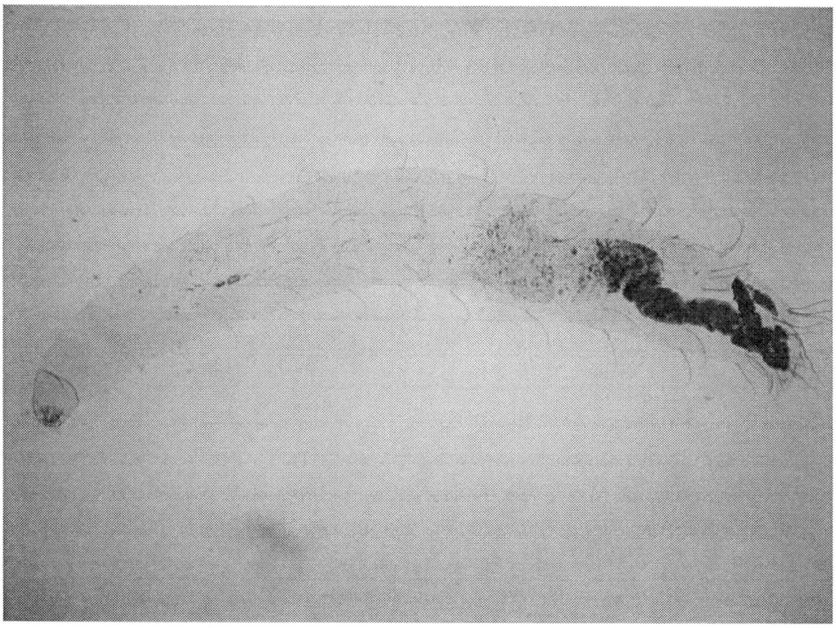

Figure 16. The hatched larvae are free-living in all three larval instars.

e. Diagnosis

Differentials include atopy, food hypersensitivity, scabies, cheyletiellosis, demodicosis and *Malassezia* dermatitis.

The diagnostic includes direct visualization of fleas or flea excreta (reddish-black faecal pellets, known as "flea dirt") on body. In suspects flea-allergic cats intradermal (reaction to flea antigen) and serological (detection of IgE antiflea antibody) allergy testing can be made. False negative results are possible (Hnilica, 2011)

f. Treatment

The control and eradication of the fleas on hosts and in the environment are the objective of treatments. Topical or systemic insect growth regulators (lufenuron, methoprene) acts on free-living stages and reduce the flea burden in the pet's environment. Products as pyrethrin, fipronil, imidacloprid, spinosad, nitenpyram, dinotefuran, metaflumizone and selamectin are efficacious when are administered (orals, spray, spot-on solutions or dips) to cats for adult flea control (Krämer and Mencke 2001; Griffin 2006; Blagburn and Dryden, 2009).

References

Arther, RG. Mites and lice: biology and control. *Veterinary Clinics Small Animal Practice,* 2009 39, 1159-1171.

Blagburn, BL & Dryden, MW. Biology, treatment, and control of flea and tick infestations. *Veterinary Clinics Small Animal Practice,* 2009 39, 1173-1200.

Desch, CE & Stewart, TB. Demodex gatoi: new species of hair follicle mite (Acari: Demodecidae) from the domestic cat (Carnivora: Felidae). *Journal of Medical Entomology,* 1999 36, 167-70.

Dryden, MW & Rust, MK. The cat flea: biology, ecology and control. *Veterinary Parasitology,* 1994 52, 1-19.

Foley, RH. Parasitic mites of dogs and cats. *Compendium on Continuing Education for the Practicing Veterinarian,* 1991 13, 297-311.

Ghubash R. Parasitic miticidal therapy. *Clinical Techniques in Small Animal Practice,* 2006 21:135-144.

Griffin, CE. Flea Allergy Dermatitis. In: Birchard, SJ & Sherding, RG, editors. Saunders Manual of Small Animal Practice. Third Edition. Elsevier; 2006.

Hnilica, KA. *Small Animal Dermatology: A Color Atlas and Therapeutic Guide*. Third edition. Elsevier; 2011.

Krämer, F & Mencke, N. Flea biology and control: The biology of the cat flea. *Control and prevention with imidacloprid in small animals.* Springer Berlin, Heidelberg,

Morgan, RV. *Handbook of Small Animal Practice.* 5th Edition.Saunders; 2007.

Paterson, S. *Manual of Skin Diseases of The Dog and Cat.* Second edition. Blackwell Publishing; 2008.

Pinchbeck, LR & Hillier, A. Scabies, Notoedric Mange and Cheyletiellosis. In: Birchard, SJ & Sherding, RG, editors. *Saunders Manual of Small Animal Practice.* Third Edition. Elsevier; 2006.

Urquhart, GM; Armour, J; Duncan, JL; Dunn, AM &Jennings, FW. *Veterinary Parasitology.* Second Edition. Blackwell Science Ltd. 1996.

Wall, R & Shearer, D. 1997. Veterinary Entomology: *Arthropod Ectoparasites of Veterinary Importance.* First edition, Chapman & Hall; 1997.

Wall, R & Shearer, D. *Veterinary Ectoparasites: Biology, Pathology and Control.* Second edition. Blackwell Science; 2001.

In: Cats
Editors: T. Zakrevsky and V. Lachinov
ISBN 978-1-62081-043-9

Chapter II

DEVELOPMENT OF CAT SYMPATHETIC GANGLIA IN POSTNATAL ONTOGENESIS

P. M. Masliukov
Department of Normal Physiology, Yaroslavl State Medical Academy,
Revolucionnaya, Yaroslavl, Russia

ABSTRACT

In cats, sympathetic ganglia divide into para- and prevertebral. The largest paravertebral ganglia are the superior cervical and stellate. The superior cervical ganglion contains neurons that supply sympathetic innervation to the face, the stellate ganglion innervatethe organs of neck and thoracic cavity. Prevertebral ganglia composed of postganglionic sympathetic principal neurons that supply abdominal and pelvic viscera. Anatomical structure of the cat sympathetic ganglia is already formed to the moment of birth.

Nevertheless, elements of the cat stellate ganglion are not mature to the moment of birth. Maturation accompanies with the increasing of the neurons size and the decreasing of their number in the ganglion, with the reorganization of the connections of neurons in the ganglion. In postnatal ontogenesis in kittens, most of connections of the SG neurons with the cells-targets of viscera are formed in the first 2-3 weeks of life.

Electrophysiological properties also changes: conduction velocity and amplitude of discharges in sympathetic nerves increases, their forms become more complicated. Rhythmical discharges in sympathetic nerves are observed from the moment of birth and correlate with cardiac (1.2-3 Hz) and respiratory frequency (0.15-0.7 Hz). Amplitude of discharges is minimal in newborn kittens and increases during ontogenesis. Relative spectral power appears to be maximal at cardiac and respiratory frequencies in animals in all age groups. Morphological and functional processes of maturation pass heterochroneously and finish in kittens to six month of life.

Introduction

Sympathetic ganglia divide into para- and prevertebral. The largest paravertebral ganglia are the superior cervical and stellate. The superior cervical ganglion contains neurons that supply sympathetic innervation to the face, the stellate ganglion innervatethe organs of neck and thoracic cavity. Prevertebral ganglia composed of postganglionic sympathetic principal neurons that supply abdominal and pelvic viscera. However, there are not enough data concerning the anatomy and physiology of the cat sympathetic ganglia and their development during the ontogenesis.

Gross Anatomy of the Cat Sympathetic Ganglia

The sympathetic trunk lies along the sides of the bodies of the vertebrae ventral to the heads of the ribs, and consists of a series of ganglia, for the most part segmentally arranged, bound together by a continuous nerve cord. Each spinal nerve is connected with an adjacent ganglion by a gray ramus. A ganglion may send a gray ramus to two or more nerves, in which case it represents a combination of two or more segmental ganglia (Jänig, 2006). Some compound ganglia are constant in their occurrence, as in the case of the stellate and the superior cervical ganglia; others may occur occasionally in any part of the trunk due to the fusion of two segmental ganglia. Thus the segmental character of the truncus is indicated by the gray rami, although in some cases adjacent segments may be fused. The white rami are much more irregular in their distribution (Skok, 1973; Gabella, 1976; Nozdrachev, 1983).

The superior cervical ganglion (SCG) is quite large and located opposite the second and third cervical vertebrae (Skok, 1973; Nozdrachev, 1983). This ganglion, which forms the cephalic end of the sympathetic trunk, has a greater variety of connections than any other ganglion in the body (Figure 1).

Its branches run to three cranial nerves, three spinal nerves, several arteries, the carotid glomus, the thyroid, salivary, and lacrimal glands, smooth muscle of the eye, the glands and vessels of the mucous membrane of the head, the skin, and the smooth muscle of the hair follicles (Nozdrachev, 1983; Hayakawa et al., 2000; Jänig, 2006). The SCG is usually fusiform in shape; it is in relation, in front, with the sheath of the internal carotid artery and internal jugular vein; behind, with the longus capitis muscle. SCG sends their branches to the first three spinal nerves. The superior cervical ganglia of cats are in close proximity to the nodose ganglia (Ranson, Billingley, 1918a). A constant communication is found with the nodose ganglion of the vagus or with the superior laryngeal branch of the vagus just after it has left the ganglion. A constant branch is found coming from the upper pole, runs upward toward the base of the skull, and is found to communicate with the twelfth nerve within its bony canal. A branch to the glossopharyngeal nerve is also constantly present. A small number of branches coming from the medial side near the upper pole runs medially over the carotid artery to the pharynx. A constant branch to the superior thyroid artery is to be found coming off from the lower one-third of the ganglion on the medial side. The external carotid plexus was seen to be formed by two large branches from about the middle of the medial side of the ganglion, which ran upward and slightly medialward to the bifurcation of the common

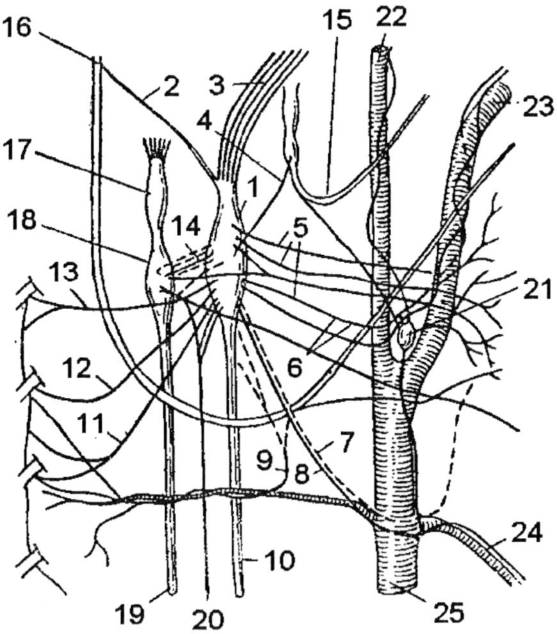

Figure 1. The branches of the superior cervical ganglion of the cat. 1- SCG, 2, 7 - branches to the hypoglossal nerve, 3 - internal carotid nerve, 4 - branch to the glossopharyngeal nerve, 5 - branches to the pharynx, 6 - branches to the external carotid artery, 8 - branches to the superior thyroid artery, 9, 11-13 - gray rami, 10 – sympathetic trunk, 14 - branches to the vagus nerve, 15 - glossopharyngeal nerve, 16 - hypoglossal nerve, 17 – jugular ganglion, 18 – nodose ganglion, 19 - vagus nerve, 20 – depressor nerve, 21 - carotid glomus, 22, 23 – internal and external carotid artery, 24 - superior thyroid artery.

carotid artery. From the upper end of the ganglion spring four or five large branches which are continued directly upward and enter the skull with the internal carotid artery and form the internal carotid nerve (Ranson, Billingley, 1918b).

The middle cervical ganglion (ganglion cervicale medium) is the smallest of the three cervical ganglia. It is in the apical portion of the thoracic cage, undoubtedly contributing to failure of some authors to identify this structure (Ranson, Billingley, 1918b). At least one major nerve connects the right middle cervical ganglion with the right thoracic vagus. The cervical sympathetic trunk, arising from the cranial pole of the right middle cervical ganglion, courses cranially to join the superior cervical ganglion; it lies lateral to the cervical vagus throughout its course. The middle cardiac nerve, arises from the middle cervical ganglion, or from the trunk between the middle and inferior ganglia. It descends behind the common carotid artery, trachea, receives a few filaments from the recurrent nerve, and joins the right half of the deep part of the cardiac plexus. In the neck, it communicates with the superior cardiac and recurrent nerves (Phillips et al, 1986).

The stellate ganglion (SG) is known as important center of sympathetic innervation of organs of the thoracic cavity, neck and upper extremity (Skok, 1973; Nozdrachev, 1983; Maslyukov et al., 2010). The SG lies ventral to the first intercostal space and the angle of the second rib. It receives the gray and white rami from the first three thoracic nerves and sometimes from the fourth also (Phillips et al, 1986; Nozdrachev et al., 2003).

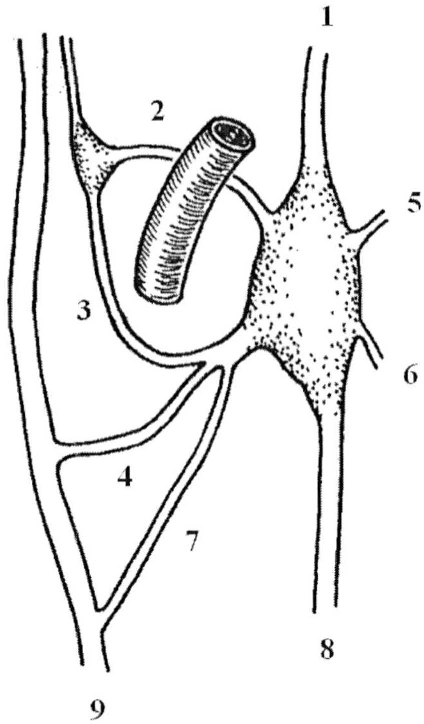

Figure 2. The nerves of the stellate ganglion. 1 - vertebral nerve; 2, 3 - cranial and caudal branches of the subclavian loop; 4 - communicating branch with the vagus nerve; 5, 6 - communicating branches T1 and T2; 7 - inferior cardiac nerve; 8 - sympathetic trunk; 9 - vago-sympathetic trunk.

The morphologic study of the SG showed, that the ganglia had different forms from typically stellar up to ellipse; extended mostly in rostrocaudal direction and slightly in mediolateral. The right SG of cats is relatively large and globular (Figure 2). It is located lateral to the longus colli muscle and over the base of the first rib, receives rami from the lower cervical and upper two thoracic segmental nerves, and connects at its caudal pole with the sympathetic chain (Skok, 1973; Gabella, 1976). A relatively large vertebral nerve, a large ventral subclavian ansa, and one or two relatively small dorsal ansae arise from the cranial region of this ganglion (Fateev, Masliukov, 1994).

A right stellate cardiac nerve arose from the medial surface of this ganglion and coursed toward the heart. The ventral ansa passes medially from the SG, frequently giving rise to one or two branches, courses cranially for several centimeters before it joins the right middle cervical ganglion, located near the thoracic outlet. The right dorsal ansa, which usually is singular but may be double, courses cranially dorsal to the subclavian artery to join the caudal pole of the right middle cervical ganglion, sometimes giving off branches which join the ventral ansa (Ranson, Billingley, 1918b; Phillips et al, 1986).

The left SG is located in the dorsolateral aspect of the cranial mediastinum, lateral to the longus colli muscle on the neck of the first and second ribs. It is globular in shape and receives rami communicans from the first and second thoracic nerves, as well as from the lower cervical nerves. A vertebral nerve arises from the cranial pole of this ganglion (Fateev, Masliukov, 1994). The sympathetic chain and a branch to the third thoracic segmental nerve arise from its caudal pole. The dorsal ansa, a small nerve arising from the medial aspect of its cranial pole, passes dorsally and cranially for 2 to 5 cm to unite with the left middle cervical ganglion near the thoracic outlet. The ventral ansa is usually larger than the dorsal ansa and sometimes consists of more than one nerve. It courses medially from the SG, ventral to the subclavian artery, before coursing cranially for 3 to 5 cm to join the left middle cervical ganglion adjacent to where the dorsal ansa joins this ganglion. The ventral lateral cardiac nerve courses ventral to the aortic arch and left pulmonary artery to join the heart adjacent to the cranial portion of the left atrial appendage (Ranson, Billingley, 1918b; Phillips et al, 1986). At least three cardiac nerves arise from the medial aspect of the left thoracic vagus. The innominate nerve originates from the cranial and medial aspect of the thoracic vagus at the level of the middle cervical ganglion. The innominate nerve courses between the common carotid and left subclavian arteries to lie adjacent to the innominate artery. At least two other nerves arise from the medial surface of the thoracic vagus above the recurrent laryngeal nerve - the ventral medial cardiac nerve and cranial vagal cardiac nerve. The ventral medial cardiac nerve usually courses toward the heart adjacent to the left subclavian artery. The left stellate cardiac nerve arose from the left stellate ganglion usually within the same sheath as the left ventral ansa. It courses caudally to the heart, lateral to the other cardiopulmonary nerves. This nerve frequently contributed tiny branches to the ventrolateral and/or the ventromedial cardiac nerves, as well as to the thoracic vagus. Cardiac nerves also originate from the left recurrent laryngeal nerve as it courses around the arch of the aorta-the left recurrent cardiac nerves. Unnamed cardiac nerves also arise from the left thoracic vagus between the level of the origin of the left recurrent laryngeal nerve and the midhilar regions of the lung (Phillips et al, 1986).

The length and width of the SG were on the right side 5.10±0.18 mm and 3.20±0.14 mm and 5.40±0.14 mm and 2.70±0.10 mm on the left. Usually, right ganglion is wider than left (Nozdrachev et al., 2003). Differences of the formation of branches in the right and left stellate ganglia were revealed. Basically they were connected with the medial pole of ganglion. In the right ganglion 3 variants were observed. The most frequent variant was one branch in the form of the anastomosis with the vagus nerve, formed by the confluence of the caudal branch of the subclavian loop and inferior cardiac nerve. In the second variant, the caudal branch of the subclavian loop and inferior cardiac nerve were formed separately. The last variant was represented by 3 branches originating from the stellate ganglion: the caudal branch of the subclavian loop, connecting branch with the vagus nerve and the inferior cardiac nerve. Instead in the left ganglion 5 variants were observed. The most frequent variant of the left SG corresponded to the third variant in the right ganglion. Beside the above mentioned, two variants were else revealed in the left ganglion. Not three, but four branches originated from the medial pole of the ganglion, last of which (accessory inferior cardiac nerve) began closely to caudal pole of the ganglion (first three branches are similar to the above named). The most rarely encountered variant was, when the caudal branch of the subclavian loop, inferior cardiac nerve and accessory inferior cardiac nerve originated directly from the stellate ganglion (Nozdrachev et al., 2003).

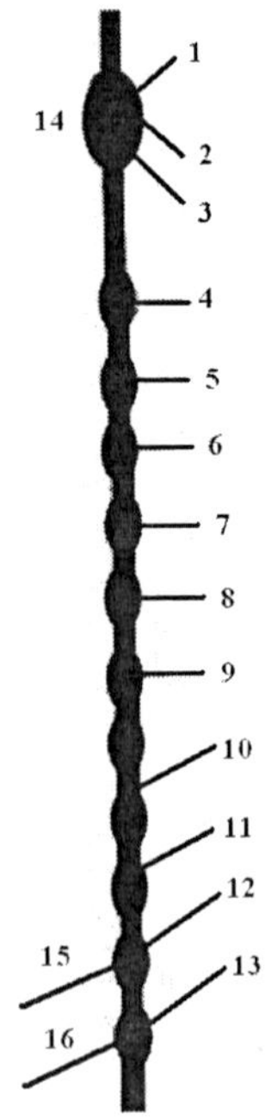

Figure 3. Thoracic portion of the sympathetic trunk in cats. 1 – 13 - communicating branches from T1 till T13, 14 – stellate ganglion, 15 - major splanchnic nerve, 16 - minor splanchnic nerve.

The thoracic portion of the sympathetic trunk consists of a series of ganglia (Figure 3). The ganglia are small in size. Two rami communicantes, a white and a gray, connect each ganglion with its corresponding spinal nerve (Jänig, 2006). Beginning with the fourth or fifth thoracic segment, the ganglia are more regularly arranged. But sometimes a ganglion from which a gray ramus arises may be very small, and occasionally scattered ganglion cells or even fairly large ganglionic masses are found in the internodal segments (Baron et al., 1985; Percy et al., 1988). The rami communicantes of the first two thoracic nerves in the cat run directly to the stellate ganglion. From the fourth to the eighth spinal nerve the gray and white rami run close together and join the trunk at the level of the corresponding ganglion. From the level of the ninth or tenth thoracic nerve to the fourth lumbar the white rami are directed caudad and reach the truncus below the level of the ganglion to which the corresponding gray rami run, often at the level of the next ganglion below (Ranson, Billingley, 1918c, d).

The two lumbosacral sympathetic chains lie close to each other on the ventral surfaces of the lumbar and sacral vertebrae. There are variations in the rostrocaudal position of the ganglion lobes relative to the vertebrae and spinal nerves. Irregularities in the distribution of ganglia were common in the upper lumbar region, in association with the major (thoracic) splanchnic nerve and other branches to the coeliac (CG) and superior mesenteric ganglia (SMG), and to a lesser extent at the lumbosacral junction (Baron et al., 1985; Percy et al., 1988). In the upper lumbar region, ganglia were sometimes split into multiple small lobes, often displaced ventrally on the crus of the diaphragm. At more caudal segmental levels (from L3 down), where the commonest position of the ganglion was immediately lateral to the lumbar spinal blood vessels, two (and rarely more) lobes were sometimes present. Less often, adjacent segmental ganglia were fused together at an intermediate location. In caudal ganglia, below the promontorium of the first sacral vertebra, fusion of ganglia or at least some gross connection between sides occurred frequently. The chains became separate again in the pelvic region. Connection between the chains was present at both L7 and S1 in the majority of cases. When the chains remained separate, L7 ganglia were situated more rostrally overlying the promontorium (Baron et al., 1985).

The greater splanchnic nerve of the cat usually leaves the trunk just below the level of the thirteenth thoracic ganglion. It is formed by the separation of a large part of the oval well myelinated fascicle from the rest of the sympathetic trunk and is composed of fine myelinated preganglionic fibers, destined to end in the coeliac ganglion, and of both myelinated and unmyelinated afferent fibers. Occasionally it also contains a bundle of postganglionic fibers arising from cells in ganglia of the sympathetic trunk (Stoyanova et al., 1997).

In the cat, the splanchnic fibers leave the sympathetic trunk in a series of six to ten small nerves. The first of this series is the largest and corresponds to the greater splanchnic nerve in man (Ranson, Billingley, 1918c, d). There were between one and five (usually two) rami projecting towards the CG and SMG from each of LI and L2 sympathetic ganglia (minor splanchnic nerves), while the lumbar splanchnic nerves to the inferior mesenteric ganglion (IMG) usually arose from L3 - L5. Visceral rami at the lower lumbar levels were detected extremely rarely; none could be found arising from sacral ganglia (Baron et al., 1985).

The great splanchnic nerve joins the CG just caudad of the celiac artery, almost on the lateral surface of the superior mesenteric. From it nerves branch in all directions, forming the cceliac plexus, a part of the large solar plexus. Less than a centimeter ventrocaudal of the CG, on the caudal surface of the superior mesenteric artery is the somewhat smaller SMG; the two are connected by two strong nerve trunks (Figure 4). The left CG, SMG, and right CG are often fused into a single complex ganglion (solar plexus, Ranson, Billingley, 1918c, d; Decktor, Weems, 1983 Fasano, Niel, 2009).

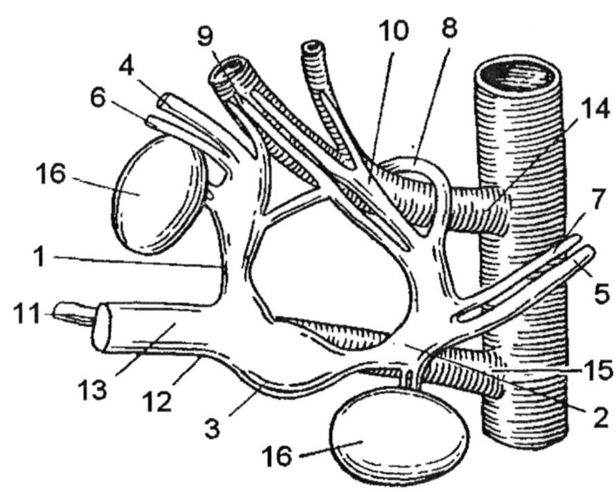

Figure 4. The celiac-superior mesenteric ganglion comlex. 1, 2 – CG, 3 - SMG, 4, 5 – major splanchnic nerve, 6, 7 - minor splanchnic nerve, 8 – interganglionic communicating branch, 9 – lateral nerve, 10 – medial nerve, 11 – superior nerve, 12 – inferior nerve, 13 – ventral nerve, 14 – celiac artery, 15 - superior mesenteric artery, 16 – suprarenal gland.

The CG receives from the sympathetic, in addition to the great splanchnic nerve, two (or three) lesser splanchnic nerves, which arise from the main trunk of the sympathetic, the first just as it reaches the abdominal cavity, the second one or two centimeters caudad of the diaphragm. From the CG and SMG a network of nerves passes in all directions to the abdominal viscera (Skok, 1973; Szurszewski, 1981; Miolan, Niel, 1996). These nerves follow especially the arteries, forming plexuses over their surfaces, and passing with them to the organs which they supply. The celiac plexus anastomoses with the posterior gastric plexus of the vagus. Parts of the celiac plexus following the branches of the celiac artery may be distinguished as hepatic plexus, splenic plexus, etc. A dense plexus is formed on the surface of the suprarenal body (suprarenal plexus), and a renal plexus follows the renal artery to the kidneys. The superior mesenteric plexus forms a dense network over the superior mesenteric artery and follows it to the intestine. An aortic plexus passes caudal on the aorta. This is a continuation of the celiac and superior mesenteric plexuses; it receives a number of branches from the main sympathetic trunk, lying dorsal of the aorta. The aortic plexus contains a number of small ganglia. Of the branches of distribution, some pass in front of the aorta, and join the aortic plexus; others descend in front of the common iliac arteries, and assist in forming the hypogastric plexus. From the aortic plexus the smaller inferior mesenteric plexus follows the inferior mesenteric artery to the large intestine; in this lies a ganglion of considerable size, the inferior mesenteric ganglion (IMG, Gabella, 1976; Nozdrachev , 1983).

The IMG usually had 4 lobes, more or less clearly defined: two proximal to the inferior mesenteric artery from which arose most or all of the colonic nerves, and two distal to this artery projecting mainly via the hypogastric nerves (Figure 5). From the rostral pole of the IMG complex, a loose bundle of connective tissue and blood vessels containing nerve strands and a number of small ganglia of variable size connected the IMG to the superior mesenteric ganglia (Krier et al., 1982). On each side of the IMG, its spinal rami (the lumbar splanchnic nerves) travel across from the lumbar sympathetic chain. From 3 to 6 nerve branches arose from the chain. There was some branching and fusion between these nerves so that the number entering the IMG was generally fewer than the number leaving the chain (range 2-5). One or two small intercalated (accessory) ganglia were sometimes located at the junction of two lumbar splanchnic nerves. These small lobes were associated in most instances with the lumbar splanchnic nerves arising from L3 and L4.

The lumbar splanchnic nerves usually originated in relation to the white rami of L2-L5 segments with the thickest nerves arising at or near ganglia L3 and L4 of the lumbar sympathetic chain. The visceral rami from L2 usually originated from the last minor splanchnic nerve (which ran ventrally and medially to the celiac and superior mesenteric ganglia) and joined the next most caudal lumbar splanchnic nerve which arose from ganglion L3. One or two thin nerves ran from the most rostral splanchnic nerve to the intermesenteric nerve or superior mesenteric or celiac ganglion (Nozdrachev, 1983; Baron et al., 1985).

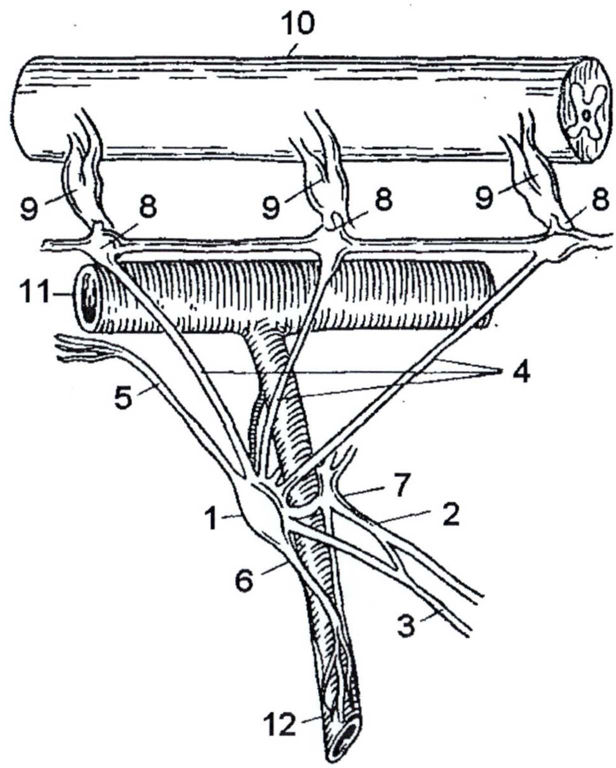

Figure 5. The inferior mesenteric ganglion and its nerves. 1 - IMG, 2, 3 – hypogastric nerve, 4 – lumbar splanchnic nerve, 5 – intermesenteric nerve, 6 - inferior mesenteric nerve, 7 - interganglionic communicating branch, 8 – ganglia of sympathetic trunk, 9 – lumbar spinal nerve, 10 – spinal cord, 11 – aorta, 12 - inferior mesenteric artery.

The sacral trunk presents many variations in macroscopic structure. As a rule, it consists of three small ganglionic enlargements. The number of ganglia, however, may be reduced or increased or the ganglion cells may be scattered throughout the length of the trunk. The sacral trunk presents many variations in macroscopic structure. As a rule, it consists of three small ganglionic enlargements. The number of ganglia, however, may be reduced or increased or the ganglion cells may be scattered throughout the length of the trunk. Adjacent ganglia are often connected by two or more strands representing the internodal segment. Fusions between opposite ganglia of the same segment are common (Johnson, 1921; Gabella, 1976).

Morphology of Sympathetic Ganglionic Neurons

The most of sympathetic neurones have well developed dendritic arborizations (Purves 1988). They usually have from 6 to 12 primary dendrites and have a total dendritic tree length exceeding 3 mm. The soma size of sympathetic neurones also varies widely, with the somatic surface area varying by a factor of two to three between neurones in the same ganglion (Michaelis et al. 1993, Gibbins et al., 2000, 2003). Nevertheless, there is no consistent correlation between the somatic size and the size of the dendritic tree. Many neurons have an asymmetric distribution of their dendrites, with virtually all of the primary dendrites arising from a single region of the soma. The dendrites branch irregularly, often in the plane of the ganglion and show little tapering (Purves. 1988; Gibbins et al., 2000, 2003).

The average area, average and maximal diameter of neurocytes bodies in the left SG (643.2±5.07 μm^2, 27.7±0.11 μm and 35.6±0.15 μm accordingly) was reliably more, than in the right SG (624.7±5.23 μm^2, 27.3±0.12 μm and 34.9±0.16 μm accordingly). As these parameters had a high degree of correlation among themselves and basically neurocytes had bodies of oval form. The maximal diameter of neuronal soma varied from 10 up to 80 μm. Most neurons had the diameter of 20-50 μm (86.8 % in the right and 87.2 % in left ganglia); 6.3% of cells in the right and 8.0% in the left SG were more than 50 μm in diameter. The number of small cells with diameter from 10 up to 20 μm was larger in the right SG and the number of large neurocytes with the diameter of 50-60 μm was predominant in the left. Neurocytes density in the right SG 654±7/mm^2 was larger than in the left 508±9/mm^2. The mean number of neurons in the SG was 376,434±21,994. The number of neurons in the right SG (467,142±27,070 cells) was more in comparison with the left SG (293,424±14,121 cells, Nozdrachev et al., 2003).

Connections of Sympathetic Neurons with Target Organs

Sympathetic ganglionic neurons innervating the target-organs are located in the certain areas of the ganglia on a topical basis (Fateev, Nozdrachev, 1995). In the SCG, neurons projecting to the iris, Muller's smooth muscle of eyelid, submandibular gland, the tongue, the mucosa of cheeks and the skin in various areas of the head and the neck were found (Morgan et al., 1987; Kuchiiwa, Kuchiiwa, 1996; Hayakawa et al., 2000). Cells of origin of sympathetic fibers innervating the extraocular muscles were found to be distributed widely to the ipsilateral SCG. Majority of these were medium in size. There were some large-sized neurons which frequently showed to have long dendrites expanding rostrocaudally (Wong, Leong, 1984). The labeling was characterized by localization in the rostral two-thirds of the SCG after injections of retrograde tracer into the medial, lateral and inferior and superior oblique muscles. However, the labeling was found in the whole of the SCG after injections into the superior rectus and inferior oblique muscles (Hayakawa et al., 2000).

Labeled neurons in the SG were localized in certain areas after injection of retrograde tracer in different organs (Fateev, Nozdrachev, 1995). After injection of tracer in the sternocleidomastoid muscle labeled neurons were detected in the zone adjacent to the emergence from the SG of the vertebral nerve (Figure 6). Labeled neurons were found in the zone of the emergence of the caudal branch of the subclavian loop and of the inferior cardiac nerve following injection of retrograde tracer into the bifurcation of the trachea, the middle third of the esophagus and ventricles of the heart. Injection of tracer into the thyroid and the cardiac part of the stomach has allowed to reveal a very small number of labeled neurons in the central part of the SG (Fateev, Nozdrachev, 1995; Nozdrachev et al., 2002, 2003).

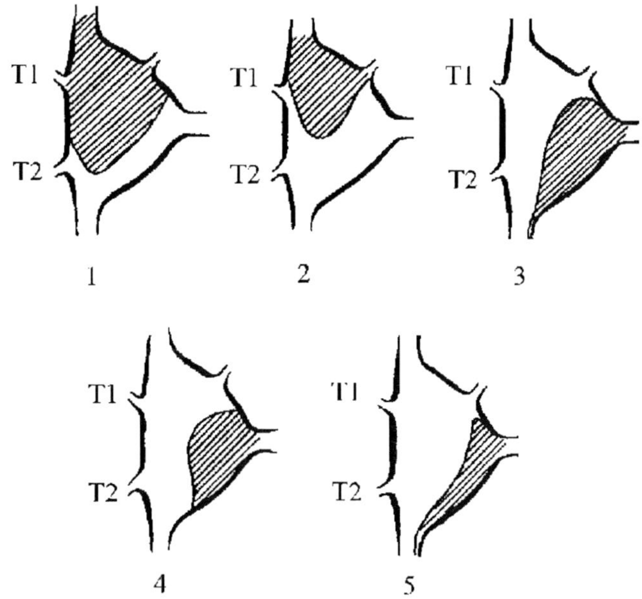

Figure 6. Scheme of location of the SG neurons innervating (1) vessels of biceps brachii muscle, (2) vessels of sternocleidomastoid muscle, (3) esophagus, (4) heart (5) trachea. T1, T2 - communicating branches with thoracic segmental nerves T1 and T2.

In the SG, 46.1 % of all labeled neurons participated in the innervation of the sternocleidomastoid muscle, 31.0% in the innervation of the trachea, 17.1% of the esophagus, 5.3% of the heart, 0.4% of the stomach and about 0.1% of the thyroid.

The largest area occupied by labeled cells was observed with HRP application to the esophagus (3.5 mm^2), then in the sternocleidomastoid muscle (3.4 mm^2), the heart (3.0 mm^2) and the trachea (2.0 mm^2), that, in our opinion, give a more objective picture. The partial overlapping was observed in the zone of the sternocleidomastoid muscle with the zones of the esophagus and the heart. The zones of the trachea, the esophagus and the heart were completely overlapped. The received results give the evidence about polyfunctionality of these zones in the supply of several at once (Fateev, Nozdrachev, 1995; Nozdrachev et al., 2002, 2003).

Neurons innervating the small intestine were located in the upper part of the SMG. Most labeled cell bodies of renal nerves were clustered in groups within the solar plexus, but some cell bodies of renal neurons were observed in upper lumbar (L1-L3) and lower thoracic (T12-T13) paravertebral sympathetic ganglia. In contrast, 90% of labeled splenic neurons were scattered randomly throughout the left and right celiac poles of the solar plexus (Meckler, Weaver, 1984).

Sympathetic postganglionic nerve cell bodies projecting in the splenic nerve of the cat were found bilaterally in sympathetic paravertebral ganglia T3-L2 with a maximum in T10-T13 and in the coeliac superior mesenteric ganglion complex. About 98% of the postganglionic neurons were located in prevertebral and 2% in paravertebral sympathetic ganglia. About 66% of the postganglionic paravertebral neurons were located on the left side (Baron, Jänig, 1988).

The sympathetic postganglionic neurons that project to the colon along the inferior mesenteric artery of the cat were labeled bilaterally in the proximal lobes of the inferior mesenteric ganglion (70-95%), in accessory ganglia of the intermesenteric nerve and of the lumbar splanchnic nerves, and in lumbar paravertebral ganglia. Labeled postganglionic cell bodies in both the IMG and the accessory ganglia were larger than labeled and unlabeled ganglion cells in the paravertebral ganglia (Baron et al., 1985).

Conduction Pathways of the Cat Sympathetic Ganglia

The conductive pathways of the cat sympathetic ganglia can be divided into three groups according to their functional features. The first group includes efferent synaptically switched pathways, the second, efferent preganglionic transient and the third, afferent spinal transient pathways.

In the SCG, 4 groups of preganglionic fibres were found on the basis of their electrophysiological properties. Preganglionic conduction velocities and synaptic delays were within the ranges 15-26 m/sec and 2.2-3.8 msec, 10-14 m/sec and 4.5-8.4 msec, 7-10 m/sec and 12-18 msec, 1.2-5.0 m/sec and 10-32 msec, respectively (Skok, 1973; Zefirov, Khirug, 1984).

In the SG, synaptically switched pathways were present in all ganglionic branches emerging from the SG except the thoracic part of the sympathetic trunk (Skok, 1973; Nozdrachev et al., 2003). Transient pathways passed through the structure of all branches in the cat SG (Skok, 1973; Nozdrachev et al., 2003). The fibers of these pathways were divided into three groups: A_δ, B and C according to their conduction velocities. The fibers of the A_δ and B groups had a mean conduction velocity from 9.2±0.82 to 12.0±2.00 m/s. These fibers passed through all branches of the SG. The fibers of the C group were divided into two subgroups on their reliably different conduction velocities. The first group consisted of the fastest fibers with a mean conduction velocity from 1.8±0.08 to 2.0±0.07 m/s. The fibers of the second subgroup had a mean conduction velocity from 1.2±0.07 to 1.4±0.06 m/s (Fateev, Nozdrachev, 1996; Nozdrachev et al., 2003).

In the SMG nerves, three types of afferent A-fibers were observed: 1) fibres with the maximum conduction velocity of 48-85 m/s; 2) fibres with the conduction velocity from 8-10 to 33-39 m/s; 3) fibres with the conduction velocity from 1.8 to 7 m/s. 4) Afferent and postganglionic C-fibres had the conduction velocity below 1.8 m/s (Lukoshkova, 1975). The conduction velocity in the major splanchnic nerve was 4.1-5.0 m/s (McAllen, May, 1990).

In the IMG, three types of fibers were identified based on their conduction velocity of excitation (Lloyd, 1937). The conduction velocity of the first two types were 12.9±1.36 and 3.3±0.39 m/s. The third type fibers had a synaptic switch in the IMG. The synaptic delay was 12.8±0.9 ms in the case of recording in ipsilateral branches and 63.1±15.5 ms in contralateral branches. The conduction velocity of excitation in these pathways was 2.85±0.16 in preganglionic fibers and 1.49±0.40 m/s in postganglionic ones (Nozdrachev et al., 1970).

SYMPATHETIC NERVOUS ACTIVITY

Adrian et al. (1932) first described the activity in sympathetic nerves in anaesthetized cats.. The discharges of sympathetic nerves with cardiovascular targets contain mixtures of centrally generated rhythms, including slow (<1 Hz) respiratory-related, rapid cardiac-related (CR; 2–4 Hz range), and 10-Hz (7–13 Hz range) rhythms (Gebber, 1976; Barman, Kenney, 2007). These rhythms are reflected by sharp peaks in the autospectrum of sympathetic nerve discharge. But not all sympathetic nerves have the cardiac-related frequency. For example, cutaneous vasoconstrictor fibres, sudomotor fibres, adrenaline-regulating adrenal preganglionic neurons and nerves supplying the brown adipose tissue are generally devoid of this rhythm. A respiratory related rhythm is rarely seen in cutaneous sympathetic fibres in the cat A 10-Hz rhythm appeared in preganglionic and postganglionic sympathetic nerve discharge of unanesthetized decerebrate or conscious cats (Cohen, Gootman, 1970; Ninomiya et al., 1990; Barman et al., 1992). This rhythm can be mixed with cardiac-related activity in baroreceptor-innervated cats and with 2- to 6-Hz activity in baroreceptor-denervated cats (Gebber, 1976; Barman, Kenney, 2007). Nevertheless, rhythmic components of activity recorded from different sympathetic nerves in cats often coexisted with considerable aperiodic ('non-rhythmic') activity. But he rhythmic discharges of the cardiac and vertebral nerves were much more strongly correlated than their aperiodic activity. This high coherence between the rhythmic discharges in two sympathetic nerves is not restricted to two nerves emanating from the same ganglion. Rather, the rhythmic discharges of post-ganglionic nerves receiving input from the upper thoracic levels are highly correlated to those of nerves receiving input from lower thoracic or lumbar levels and the discharges of nerves on the left and right side were also strongly correlated (Kocsis et al., 1990; Gebber et al., 1995; Barman, Kenney, 2007). Multiple levels of the central nervous system (from the forebrain to the spinal cord) contain neurons that affect sympathetic rhythmicity (Barman, Kenney, 2007; Gilbey, 2007).

DEVELOPMENT OF ANATOMY OF SYMPATHETIC GANGLIA

The development of the paravertebral ganglions in postnatal period is characterized by an increase of their mass, whereas the main structure of the ganglion by the moment of birth is already formed. During postnatal ontogeny, increases in the length and weight of the animal are consistently accompanied by increases in the size and volume of the sympathetic ganglia. Our data show that from birth to age six months the length of the ganglion increased 1.9-fold, width 1.5-fold, thickness three-fold, and volume 10-fold. The shape of SG was stellate in newborn and older kittens. Distribution of the main nerves originated from sympathetic ganglia in newborn animals were close to these parameters in adult animals (Nozdrachev, Masliukov, 2003).

Development of Morphological Properties of Sympathetic Neurons

In postnatal ontogenesis, neural elements of sympathetic ganglia pass some certain stages of the development. It accompanies morphological and functional reorganization (Masliukov 2000 a, b). Neuronal size increases in postnatal ontogenesis and neurotransmitter synthesis changes (Masliukov, Timmermans, 2004; Novakovskaia et al., 2011). The cross-sectional area of neurons in the SG and their maximal diameter increased in postnatal ontogenesis. In newborn kittens these parameters were 132.2±0.87 μm^2 and 17.0±0.07 μm, in 10-day-old - 224.6±1.81 μm^2 and 22.0±0.08 μm, in 20-day-old -337.5±2.86 μm^2 and 26.2±0.11 μm, in one-month-old - 401.8±2.80 μm^2 and 29.6±0.12 μm, in two-month-old - 465.1±3.48 μm^2 and 30.0±0.12 μm accordingly. When the same groups of neurons in the left and right SG were compared on these parameters only few statistically significant differences were observed in every age group (Masliukov, 2001).

In parallel to increasing of the average diameters, the number of small neurons decreased and the percentage of large neurons increased. This process occured most rapidly in first ten days of life when the percentage of small neurons with the cross-sectional area till 200 μm^2 reduced practically twice. In newborn kittens, neurons with the cross-sectional area larger than 600 μm^2 were absent and they first appeared in 10-day-old animals. In the SG of 10-day-old kittens, all groups of neurons (till 1200 μm^2) observed in older animals were present. It's interesting that the percentage of neurons with the area of 400-600 μm^2 was practically equal in 20-day-old (23.4%), 30-day-old (24.6%) and two-month-old (25.8%) kittens. From 20 days till two months of age, the percentage of neurons with the area of 0-200 and 200-400 μm^2 reduced and the part of cells with this parameters more than 600 μm^2 increased (Masliukov, 2001).

In parallel to the increase of neuronal sizes, the distribution density of neurons decreased and in newborn kittens it was 2994±30/mm^2, in 10-day-old - 1770±12/mm^2, in 20-day-old - 1391±8/mm^2, in one-month-old - 1071±12/mm^2, in two-month-old - 821±12/mm^2 accordingly. But in newborn, 10-day-old and 20-day-old kittens this parameter in different areas of the same SG and in the same areas of the right and left ganglia were not differ. In one-month-old kittens the average distribution density of neurons and distributions in all studied areas was more in the right ganglion relatively to the left one (Masliukov, 2001).

The number of neurons in the ganglion was maximal in newborn ($2.11 \pm 0.13 \times 10^5$) and 10-day-old kittens ($1.86 \pm 0.15 \times 10^5$). Then, the number of neurons in the SG reduced and became the same as in adult cats at 20 days of age ($1.38 \pm 0.11 \times 10^5$). In newborn kittens, the mean number of neurons in the left SG was greater in comparison with the number of cells in the right ganglion. In contrast, in animals of other ages this parameter was more in the right SG. However, in other kittens these differences were not statistically significant (Masliukov, 2001).

In the IMG, heterogeneity of the maturation of neurons and their endings was observed. In newborn kittens, mature neurons, glial cells and synapses were observed together with many immature ones. In 14-day-old animals, the proportion of immature neurons decreased, but destruction of neurons was observed more frequently. In the sympathetic ganglia of all animals, axodendritic synapses were found more frequently and axosomatic synapses were observed rarely. Finally, the ultrastructure of the inferior mesenteric ganglion in kittens becomes similar to adult animals at the end of the second month of life (Novakovskaia et al., 2011).

Development of Connection of Sympathetic Neurons with Target-Organs

Even in newborn kittens, neurons participating in innervation of specific organs or vessels of skeletal muscles were located in discrete zones of the ganglion. These neurons were found in discrete zones in the same relative position of the SG in newborn kittens and older animals, therefore, did not change their location in postnatal ontogenesis. Neurons innervating the sternocleidomastoid muscle were located in the zone adjacent to the emergence of the vertebral nerve from the ganglion. Neurons, innervating the esophagus, trachea, heart were observed in the area of the anastomosis of the SG with the vagosympathetic trunk. Neurons sending their axons to the biceps brachii muscle were located primarily in throughout the area of the central sections except the area of the sympathetic trunk. Thus, the part of area, occupied by labeled neurons on central sections was maximal when HRP was injected in the biceps in animals from newborn to adult. But the areas of distibution of neurons with axons passing to the sternocleidomastoid muscle, biceps brachii muscle and heart were the largest in 10-days and 20-days old kittens in comparison to newborn and 30-days kittens (Masliukov et al., 2000b).

In postnatal ontogenesis, the neuronal organization of the SG changes in parallel with the increasing of sizes of neurons and area of the ganglion section occupied by labeled cells. The maximal number of neurons in the SG sending their fibers to the organs of the neck, thorax and upper extremity were observed in 10-days and 20-days old kittens in comparison to newborn kittens, 30-days kittens and adult cats. In newborn kittens, neurons with the largest size participated in the heart innervation. In 10- and 20-day-old kittens, the largest neurons sent their axons to the vessels of the sternocleidomastoid muscle. But the largest part of big neurons (30-50 μm) took place in innervation of the vessels of this muscle and the heart (3.2% for the heart and 14.0% for the muscle at 10-days, 3.8% and 15.8% at 20-days kittens) accordingly to other organs (0.4-1.3% at 10-days, 0.6-1.9% at 20-days). Also in 20-days kittens quite a large proportion of neurons sent their axons to the biceps brachii muscle (6.8%). In 30-day-old kittens, the largest neurons innervated the heart, sternocleidomastoid muscle and biceps brachii muscle. The largest part of big neurons (30-50 μm) sent their fibers to the these muscles and the heart (31.5% for the heart, 40.7% for the sternocleidomastoid muscle, 33.6% for the biceps brachii muscle) in comparison to other organs (8.5-22.5%).

Among the neurons innervating the sternocleidomastoid muscle, biceps and heart, the greatest number of cells was found at 20-days. The number of neurons sending their axons to the trachea and esophagus (except their thoracic part) was less in 30-days kittens in comparison to kittens of earlier ages. Also in 10- and 20-days kittens, the number of neurons, connected with the cervical and thoracic part of the esophagus was significantly more in comparison to newborns. These data are in agreement with results of Vidovic and Hill (1988), pointing to the increasing of number of neurons, innervating the eye in the rat superior cervical ganglion between 14 and 30 days of life. It is evident that in postnatal ontogenesis in kittens most of connections of the SG neurons with the cells-targets of viscera are formed in the first 2-3 weeks of life.

Development of the Conductive Pathways in the Cat Stellate Ganglion

In newborn kittens, stimulation of the sympathetic trunk caused responses in all the studied branches. The evoked responses in the sympathetic trunk itself also were found when all branches, except the vertebral nerve, were stimulated. These responses were monophasic and homogenous in their character (Figure 7). At this age, the mean of the latent period was 14.26±0.12 ms and the amplitude was 15.94±0.38 mV. The calculated rate of the conduction velocity was 0.70±0.01 m/s. In 10-day-old kittens, the responses in the SG branches after stimulation of the sympathetic trunk and in the sympathetic trunk itself when all other nerves, except the vertebral nerve, were

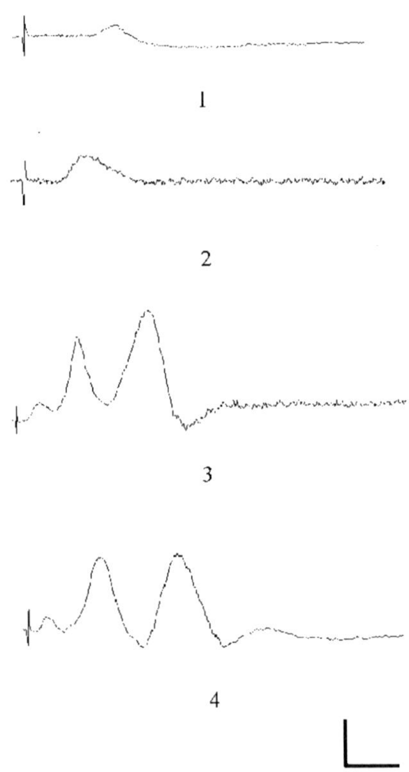

Figure 7. Evoked responses in the inferior cardiac nerve of the SG when the thoracic part of the sympathetic trunk was stimulated (1) in newborn kittens, (2) 10-day-old kittens, (3) 20-day-old kittens, (4) 30-day-old kittens.

stimulated were also monophasic. No differences were found between the latent periods of responses in this group. The means of basic electrophysiologic parameters were: latent period, 10.06±0.18 ms, amplitude, 24.92±1.64 mV, the rate of the conduction velocity of excitation, 1.02±0.02 m/s (Masliukov et al., 1999).

The evoked responses in 20-day-old kittens were more complicated than in kittens of earlier age and were divided into two groups. In the first group, the responses had a short latent period, the excitation was expanded in both directions. It testified to the presence of the transitional pathways. The responses of the second group were synaptically switched, had a longer latent period, disappeared when the irritating and recording electrode were interchanged, and also, in distinction from the first group, were eliminated by ganglioblockers. In these animals, the stimulation of the sympathetic trunk caused four-phase responses in the vertebral nerve, the caudal branch of the subclavial loop and the anastomosis with the vagosympathetic trunk, three-phase responses in the cranial branch of the subclavial loop, the connecting branch with the vagus nerve, the inferior cardiac nerve. In the sympathetic trunk itself only the biphase responses were registered after stimulation of the other branches. Since these responses were recorded in both directions, they did not disappear after using of ganglioblockers. They characterize conduction of excitation on transitional SG pathways. In the all nerves under study, the transitional pathways consisting from A_{δ}. B and C fibers were detected. The conduction velocity of excitation on fibers of groups A_{δ} and B was 4.59±0.10 m/s. Among the transitional fibers of the C group, two groups of fibers were found, authentically various on the average conduction velocity of excitation. The first consisted of fibers of the anastomosis with the vagosympathetic trunk (the average rate of conduction velocity. 1.88±0.14 m/s). In the second group, the fibers of the cranial and caudal branch of the subclavian loop, the connecting branch with the vagus nerve and the inferior cardiac nerve (the average rate of conduction velocity. 1.37±0.05 m/s) were included. In 20-day-old kittens, the average amplitude of the first phase of the responses was authentically less in comparison with the mean of amplitudes of other phases, that, probably, reflects participation of a rather small part of myelinated fibers in this age in conduction of excitation in comparison with all mass of fibers (Masliukov et al., 2000a).

Sympathetic Nervous Activity during the Development

Rhythmical discharges in sympathetic nerve activity of cats are observed from the moment of birth in different nerves (Sica et al., 1990; Masliukov, 2003; Maslyukov et al., 2010). Amplitude of discharges was minimal in newborn kittens and increased during ontogenesis.

Relative spectral power in the vertebral nerve of the SG was maximal at frequencies of 0.15-0.7 (respiratory rhythm) and 1.2-3 Hz (cardiac rhythm) in animals of all age groups (Figure 8). We did not observe statistically significant differences concerning the cardiac and respiratory rhythm in kittens of different ages. “10-Hz discharges” (8-12 Hz band) in the

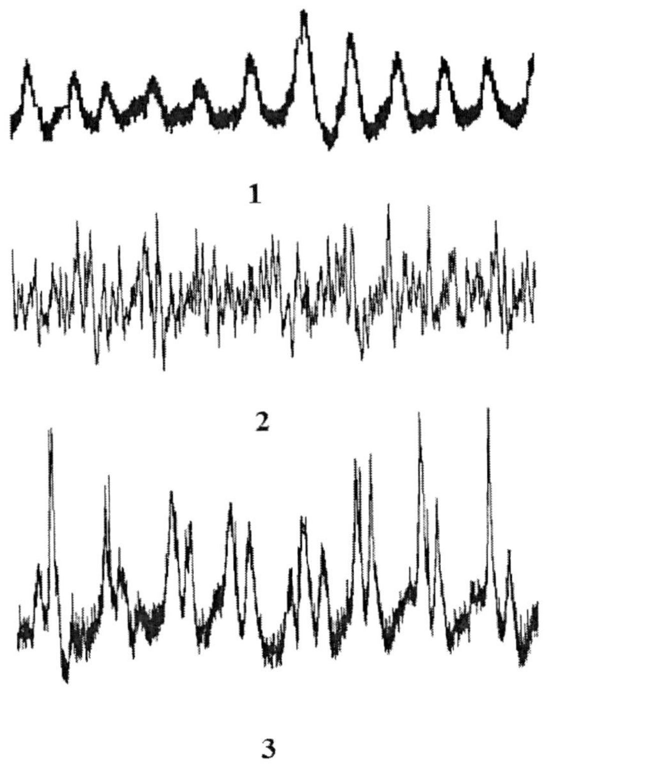

Figure 8. Discharges of the vertebral nerve of the SG in (1) newborn, (2) 30-day-old, (3) two-month-old kittens. Horizontal and vertical calibrations are 60 ms and 50 μV.

VNA appeared only in animals older than 20 days of life. In the autospectra of 20-day-old and 30-day-old kittens, we also found well-expressed frequencies of 12-40 Hz. This part of spectrum in two-month-old and older animals was absent. The spectral characteristic of discharges in two-month-old and six-month-old kittens includes 0.15-0.7, 2-3, 6-8 and 9-10 Hz (Masliukov, 2003).

In the cervical sympathetic trunk, the maximum spectral power in all kittens was at frequencies associated with the cardiac (3.2–4.2 Hz) and respiratory (0.7–0.9 Hz) rhythms. There were no significant differences in the frequency power spectra of the respiratory and cardiac rhythms within each age group or between these measures in kittens of different age groups (Maslyukov et al., 2010). The modulation by both sympathetic and inspiratory rhythm generators at specific frequencies (4-5 Hz and 8-9 Hz, respectively) was observed (Sica et al., 1990). 2-6 Hz oscillations were observed in newborn kittens, 7-10 Hz oscillations were found only in a few cats older than 4 weeks (Sica, Siddiqi, 1993).

The presence of coherent frequencies between right and left cervical sympathetic trunk discharges even in our youngest animal was indicated. In contrast, coherencies between sympathetic and phrenic nerve signals were observed only in cats older than 45 days.

CONCLUSION

To the moment of birth, sympathetic ganglia in kittens are not mature. An increase of sympathetic neuronal soma sizes, development of dendrite tree, and reorganization of nerve connections occurs during the development. A heterogeneous development of individual populations of neurons is observed. Myelination of the fibers accompanied by an increase of the excitation conduction rate and complication of the evoked response shape occurs after birth. Conduction velocity and amplitude of discharges in sympathetic nerves increases, their forms become more complicated. Amplitude of discharges is minimal in newborn kittens and increases during ontogenesis. Morphological and functional processes of maturation pass heterochroneously and finish in kittens to six month of life.

This work was supported by RFBR, a grant of the President of the Russian Federation for young scientists and the Russian Federal Special-Purpose Program "Scientific and scientific-pedagogical personnel of innovative Russia" 2009-2013.

REFERENCES

[1] Adrian, ED, Bronk, DW, Phillips, G. Discharges in mammalian sympathetic nerves. *J. Physiol.*, 1932 74, 155–133.

[2] Barman, SM., Gebber, GL., Zhong, S. The 10-Hz rhythm in sympathetic nerve discharge. *Am. J. Physiol.*, 1992 262, R1006–R1014.

[3] Barman, SM, Kenney, MJ. Methods of analysis and physiological relevance of rhythms in sympathetic nerve discharge. *Clin. Exp. Pharmacol. Physiol.*, 2007 34, 350-355.

[4] Baron, R; Jänig, W, McLachlan, EM. On the anatomical organization of the lumbosacral sympathetic chain and the lumbar splanchnic nerves of the cat - Langley revisited. *J. Auton. Nerv. Syst.*, 1985 12, 289-300.

[5] Baron, R; Jänig, W, McLachlan, EM. The afferent and sympathetic components of the lumbar spinal outflow to the colon and pelvic organs in the cat. III. The colonic nerves, incorporating an analysis of all components of the lumbar prevertebral outflow. *J. Comp. Neurol.* 1985, 158-168.

[6] Baron, R; Jänig, W. Sympathetic and afferent neurons projecting in the splenic nerve of the cat. *Neurosci. Lett.*, 1988, 109-13.

[7] Cohen, MI, Gootman, PM. Periodicities in efferent discharge of splanchnic nerve of the cat. *Am. J. Physiol.*, 1970 218, 1092-1101.

[8] Decktor, DL, Weems, WA. An intracellular characterization of neurones and neural connexions within the left coeliac ganglion of cats. *J. Physiol.*, 1983 341, 197-211.

[9] Fasano, C, Niel, JP. The mammalian sympathetic prevertebral ganglia: models for the study of neuronal networks and basic neuronal properties. *Auton Neurosci.*, 2009 150, 8-20.

[10] Fateev, MM, Masliukov, PM. The origins of the formation of the initial portion of the vertebral nerve in the cat. *Morfologiia*, 1994 107, 62-66.

[11] Fateev, MM, Nozdrachev, AD. Cat stellate ganglion pathways. *Dokl. Akad. Nauk*, 1996 351, 834-836.

[12] Fateev, MM, Nozdrachev, AD. Projections of stellate ganglion sympathetic neutrons in cats. *J. Auton. Nerv. Syst.*, 1995 51, 129-134.

[13] Gabella, G. *Structure of the autonomic nervous system.* London: Chapman and Hall; 1976.

[14] Gebber, GL. Basis for phase relations between baroreceptor and sympathetic nervous discharge. *Am. J. Physiol.*, 1976 230, 263–270.

[15] Gebber, GL, Zhong, S, Barman, SM. Synchronization of cardiac-related discharges of sympathetic nerves with inputs from widely separated spinal segments. *Am. J. Physiol.*, 1995 268, R1472–1483.

[16] Gebber, GL, Zhong, S, Paitel, Y. Bispectral analysis of complex patterns of sympathetic nerve discharge. *Am. J. Physiol.*, 1996 271, R1173–R1185.

[17] Gibbins, IL; Jobling, P; Morris, JL. Functional organization of peripheral vasomotor pathways. *Acta Physiol. Scand.,* 2003 177, 237-245.

[18] Gibbins, IL, Jobling, P, Messenger, JP, Teo EH, Morris JL. Neuronal morphology and the synaptic organisation of sympathetic ganglia // *J. Aut. Nerv. Syst.*, 2000 81, 104-109.

[19] Gilbey, MP. Sympathetic rhythms and nervous integration. *Clin. Exp. Pharmacol. Physiol.*, 2007 34, 356-361.

[20] Jänig, W. *The integrative action of the autonomic nervous system: neurobiology of homeostasis*. Cambridge University Press; 2006.

[21] Jänig, W, Sundlof, G, Wallin, BG. Discharge patterns of sympathetic neurons supplying skeletal muscle and skin in man and cat. *J. Auton. Nerv. Syst.*, 1983 7, 239–256.

[22] Johnson, SE. An experimental study of the sacral sympathetic trunk of the cat with special reference to the occurrence of intrinsic commissural neurons. *J. Comp. Neurol,* 1921 33, 85–104.

[23] Hayakawa, T, Itoh, M, Miki, T, Kaneto, T, Tomiyama, H, Takeuchi, Y. Sympathetic fibers innervating the extraocular muscles: cells of origin in the cat superior cervical ganglion. *Okajimas Folia Anat. Jpn*, 2000 77, 119-24.

[24] Kocsis, B, Gebber, GL, Barman, SM, Kenney, MJ. Relationships between activity of sympathetic nerve pairs: Phase and coherence. *Am. J. Physiol.*, 1990 259, R549–560.

[25] Krier, J, Schmalz, PF, Szurszewski, JH. Central innervation of neurones in the inferior mesenteric ganglion and of the large intestine of the cat. *J. Physiol.*, 1982 332, 125-138.

[26] Kuchiiwa, S, Kuchiiwa, T. Autonomic and sensory innervation of cat molar gland and blood vessels in the lower lip, gingiva and cheek. *J. Auton. Nerv. Syst.*, 1996 61, 227-34.

[27] Langley, JN. On the question of commissural fibers between nerve cells having the same function and situated in the same sympathetic ganglion, and on the function of post-ganglionic nerve plexuses. *J. Physiol.*, 1904 31, 244-259.

[28] Lloyd, DPC. The transmission of impulses through the inferior mesenteric ganglia. *J. Physiol.*, 1937 91, 296-313.

[29] Lukoshkova, EV. Conduction velocity and excitability of A and C fibers of the mesenteric nerve in the cat. *Neirofiziologiia*, 1975 7, 272-278.

[30] Masliukov, PM. Sympathetic neurons of the cat stellate ganglion in postnatal ontogenesis: morphometric analysis. *Auton Neurosci,* 2001 89, 48-53.

[31] Masliukov, PM. Discharge pettern of the sympathetic vertebral nerve activity in kittens in postnatal ontogenesis. *Neurosci. Lett.*, 2003 344, 141-143.

[32] Masliukov P.M., Timmermans J.-P. Immunocytochemical properties of stellate ganglion neurons during early postnatal development. *Histochem Cell Biol*, 2004 122, 201-209.

[33] Masliukov, PM, Fateev, MM, Volovenko, VN. The electrophysiological properties of the conduction pathways of the stellate ganglion in the cat in postnatal ontogeny. *Ross Fiziol. Zh Im I M Sechenova*, 1999 85, 419-429.

[34] Masliukov, PM, Fateev, MM, Nozdrachev, AD. Age-dependent changes of electrophysiologic characteristics of the stellate ganglion conducting pathways in kittens. *Auton. Neurosci,* 2000a 83, 12-18.

[35] Masliukov, PM, Pankov, VA, Strelkov, AA, Masliukova, EA, Shilkin, VV, Nozdrachev, AD. Morphological features of neurons innervating different viscera in the cat stellate ganglion in postnatal ontogenesis. *Auton. Neurosci.,* 2000b 84, 169-175.

[36] Maslyukov, PM, Korzina, MB, Emanuilov, AI. Age-related changes in rhythmic electrical activity in the cervical sympathetic trunk in rats and cats. *Neurosci. Behav. Physiol*, 2010 40, 251-256.

[37] Maslyukov, PM, Korzina, MB, Emanuilov, AI, Shilkin, VV. Neurotransmitter composition of neurons in the cranial cervical and celiac sympathetic ganglia in postnatal ontogenesis. *Neurosci. Behav. Physiol.*, 2010 40, 143-147.

[38] McAllen, RM, May, CN. The conduction velocity of the descending spinal pathway to the renal sympathetic nerve in the cat. *J. Auton. Nerv. Syst.*, 1990 30, 139-142.

[39] Meckler, RL, Weaver, LC. Comparison of the distributions of renal and splenic neurons in sympathetic ganglia. *J. Auton. Nerv. Syst..* 1984 11, 189-200.

[40] Michaelis, M, Boczek-Funcke, A, Habler, HJ, Jänig, W. Responses of lumbar vasoconstrictor neurons supplying different vascular beds to graded baroreceptor stimuli in the cat. *J. Auton. Nerv. Syst.,* 1993 42, 241–249.

[41] Miolan, JP, Niel, JP. The mammalian sympathetic prevertebral ganglia: integrative properties and role in the nervous control of digestive tract motility. *J. Auton. Nerv. Syst.*, 1996 58, 125-138.

[42] Morgan, C, DeGroat, WC, Jannetta, PJ. Sympathetic innervation of the cornea from the superior cervical ganglion. An HRP study in the cat. *J. Auton. Nerv. Syst.*, 1987 20, 179-183.

[43] Ninomiya, I, Akiyama, T, Nishiura, N. Mechanism of cardiac-related synchronized cardiac sympathetic nerve activity in awake cats. *Am. J. Physiol.*, 1990 259, R499-R506.

[44] Nozdrachev, AD. *Physiology of vegetative nervous system.* Leningrad: Medizina ; 1983 (Russian).

[45] Nozdrachev, AD, Masliukov, PM. Neuronal organization of the stellate ganglion in mammals during postnatal ontogenesis. *Zh. Evol. Biokhim. Fiziol.* 2003 39, 190-194.

[46] Nozdrachev, AD, Bezenkina, GI, Efimova, NI. Conduction pathways of the cat caudal mesenteric sympathetic ganglion. *Fiziol. Zh SSSR Im I M Sechenova*, 1970 56, 543-551.

[47] Novakovskaia, SA, Archakova, LI, Masliukov, PM. Ultrastructure of the caudal mesenteric ganglion neurons during early development in kittens. *Morfologiia*. 2011 139, 38-42. Russian.

[48] Nozdrachev, AD, Fateev, MM, Jiménez, B, Morales, MA. Circuits and projections of cat stellate ganglion. *Arch. Med. Res.*, 2003 34, 106-115.

[49] Nozdrachev, AD, Jiménez, B, Morales, MA, Fateev, MM. Neuronal organization and cell interactions of the cat stellate ganglion. *Auton. Neurosci.,* 2002 95, 43-56.

[50] Percy, WH, Walsh, J, Krier, J. Morphological and electrophysiological properties of cat lumbar paravertebral neurones. *J. Auton. Nerv. Syst.*, 1988 24, 183-192.

[51] Purves, D. *Body and Brain*. Cambridge, MA: Harvard University Press, 1988.

[52] Ranson, SW, Billingley, PR. The superior cervical ganglion and the cervical portion of the sympathetic trunk. *J. Comp. Neurol,* 1918a 29, 313–358.

[53] Ranson, SW, Billingley, PR. Branches of the ganglion cervicale superius. *J. Comp. Neurol,* 1918b 29, 367–384.

[54] Ranson, SW, Billingley, PR. The thoracic truncus sympathicus, rami communicantes and splanchnic nerves in the cat. *J. Comp. Neurol,* 1918c 29, 405–439.

[55] Ranson, SW, Billingley, PR. An experimental analysis of the sympathetic trunk and greater splanchnic nerve in the cat. *J. Comp. Neurol,* 1918d 29, 441–456.

[56] Sica, AL, Siddiqi, ZA. Respiration-related features of sympathetic discharges in the developing kitten. *J. Auton. Nerv. Syst.*, 1993 44, 77-84.

[57] Sica, AL, Siddiqi, ZA, Gandhi, MR, Condemi, G. Evidence for central patterning of sympathetic discharge in kittens. *Brain Res.*, 1990 530, 349-352.

[58] Skok, VI. *Physiology of autonomic ganglia*. Tokyo: Igaku Shoin; 1973.

[59] Stoyanova, I, Chouchkov, C, Scheuermann, DW. Immunocytochemical localization of the neurons in the superior mesenteric ganglion innervating the small intestine of the cat. *Ann. Anat.*, 1997 179, 517-523.

[60] Szurszewski, JH. Physiology of mammalian prevertebral ganglia. *Annu. Rev. Physiol.,* 1981 43, 53-68.

[61] Vidovic, M, Hill, CE. Withdrawal of collaterals of sympathetic axons to the rat eye during postnatal development: the role of function. *J. Auton. Nerv. Syst.*, 1988 22, 57-65.

[62] Wong, WC, Leong, SK. The origin of postganglionic and sensory axons in the cervical sympathetic trunk of the cat: a horseradish peroxidase study. *J. Comp. Neurol*, 1984 222, 161-165.

[63] Zefirov, LN, Khirug, SS. Organization of pathways in the cranial cervical sympathetic ganglion of the cat. *Fiziol. Zh SSSR Im I M Sechenova*, 1984 70, 779-788.

In: Cats
Editors: T. Zakrevsky and V. Lachinov

ISBN 978-1-62081-043-9

Chapter III

Feline Perioperative Analgesia in a Surgeon's and Anesthesiologist's Perspective

Lysimachos G. Papazoglou, Georgios M. Kazakos and Dimitrios Raptopoulos
Department of Clinical Sciences, Clinic of Companion Animals,
Faculty of Veterinary Medicine,
Aristotle University of Thessaloniki, Thessaloniki, Greece

Abstract

Perioperative pain control has been increasingly implemented nowadays in cats undergoing various surgical procedures. However, effective analgesia provided in cats undergoing surgery is complicated because of the difficulties in pain recognition, unfamiliarity with analgesic techniques and drugs, limited availability of analgesics such as opioids, and lack of published studies evaluating the effectiveness of these drugs in the feline patients. Ovariohysterectomy, castration, onychectomy, limb amputations, thoracic, oncologic and reconstructive surgeries are among surgical procedures that require acute or chronic pain control. Perioperative multimodal and preemptive analgesia may include opioids, non-steroidal anti-inflammatory drugs (NSAIDs), α_2 agonists, ketamine and regional blocks. They may all provide safe analgesia, taking into account the differences in metabolism between cats and other species. Apart from routine pharmaceutical treatment novel surgical techniques have been devised to achieve pain relief in cats. This article reviews the physiology and pathophysiology of pain in cats undergoing minor and major surgical procedures and discusses pain diagnosis and management strategies for pain control in the feline patient. The effect of surgical technique on feline pain control is also reviewed.

INTRODUCTION

Advances in feline surgery during the past few decades are keeping pace with its canine counterpart making this area of surgery an important and exciting part of veterinary surgical care and allowing a variety of surgical procedures to be available for cats nowadays. Many of these surgical procedures are performed solely in cats to provide treatment for conditions that are unique to this species. The tremendous growth in feline surgery allowed veterinarians to perform complicated surgeries in order to achieve prolonged survival and improve the quality of life of their patients (Lamont 2002). Ovariohysterectomy, castration, onychectomy, limb amputations, and thoracic, oncologic and reconstructive surgeries are among surgical procedures commonly performed in cats. Cats are considered as pleasing surgical patients having small size, weight less than dogs and adapt well to a surgical stimulus.

Cat population outnumbers dogs in many countries of the world including US, China, Russia and in a few European including France and Italy. In US cat population is 76 million cats and in China 53 million cats in contrast to canine population which is 61 and 22 million, respectively (www.mapsofworld.com 2006). The majority of these cats are spayed, castrated or declawed so a large part of the feline population may undergo surgery at least once in its life span and therefore experiencing some type of surgical or inflammatory pain expressing an increased need for perioperative analgesia. In addition neuropathic pain associated with chronicity may be also encountered in cats although less commonly. However, effective analgesia provided in cats undergoing surgery is complicated because of the difficulties in pain recognition, unfamiliarity with analgesic techniques and drugs, limited availability of analgesics such as opioids, unique metabolism of analgesics and lack of published studies evaluating the effectiveness of these drugs in the feline patients.

Perioperative pain control has been increasingly implemented nowadays in cats undergoing various surgical procedures but there are still under treated for pain compared to dogs; recent studies published the last 10 years showed an increased trend in analgesic use in cats among, British, Canadian, French, Finish, New Zealand and South African veterinarians (Lascelles et al. 1999, Raekallio et al. 2003, Hugonard et al. 2004, Williams et al. 2005, Hewson et al. 2006, Joubert 2006). More specifically veterinarians from various parts of the world filled questionnaires for evaluation of their attitudes in perioperative pain control in cats. British and Finish practitioners although scored perioperative pain between dogs and cats similarly, large differences in the use of analgesia between dogs and cats undergoing similar surgical procedures were reported (Lascelles et al. 1999, Raekallio et al. 2003). The use of analgesic drugs in cats having surgery increased from 13% to 94% over a 5 year period in South Africa (Joubert 2006) and a similar trend was reported for Canadian practitioners (Hewson et al. 2006). In a French study although the level of interest in analgesia appeared to be equivalent to that reported in English-speaking countries provision of analgesia in both cats and dogs was considered irrational and practitioner's attitude to the use of analgesic drugs was in contrast with that of English-speaking countries; however, this attitude may change in the near future (Hugonard et al. 2004). In a similar study in New Zealand 50% of the cats undergoing castration received analgesic therapy in contrast to 65% of the dogs and 60% of cats having ovariohysterectomy were provided with analgesia in contrast to 74% of dogs (Williams et al. 2003). Over the past few years a better understanding of the pathophysiology of feline pain along with publication of clinical and experimental studies as well as review articles allowed a considerable progress in pain management in feline patients.

PAIN PHYSIOLOGY AND PATHOPHYSIOLOGY

Free nerve endings are the main pain receptors. They convert the painful stimulus to nerve impulse, which is transmitted via afferent sensory nerves. The impulse, enters the dorsal horn of the spinal cord through the dorsal root ganglia where is amended and from there is transferred, through afferent sensory tracks (e.g. spinothalamic), to the somatosensory cortex of the brain after passing from thalamus. There, it is realized from the animal as pain. The painful stimuli cause changes in both nerve endings (peripheral sensitization) and spinal cord (central sensitization), which are responsible for phenomena such as primary and secondary hyperalgesia, and allodynia (Lamont et al. 2000, Livingston and Champers 2000, Muir 2002, Raptopoulos et al. 2002). To understand the pain the animal must be conscious. During anesthesia, while the animal is not experiencing pain, the impulses still transfer in the way just described and thus can cause an increase in cardiac and / or respiratory frequency and in blood pressure. Also, centrally transferred stimuli may cause a peripheral and central sensitization, which result in increased pain after the end of anesthesia. Moreover, if the ongoing postsurgical pain associated with wound inflammation is added, it is clear that the treatment of pre- and postoperative pain will significantly improve the overall condition of the animal. Effective perioperative analgesia aims at pain relief at the immediate postoperative period, avoiding central sensitization and minimizing long-term pain (Hellyer and Fails 2003). It is reported that postoperative analgesia is superior when the anesthetic protocol include analgesics (e.g. medetomidine and ketamine) when compared to one without drugs with analgesic activity (Slingsby et al. 1998). Preoperative administration of NSAIDs further enhances analgesia. Sensitization prevention is a key goal of pain management before the challenge of painful stimuli and is called preemptive analgesia (Balmer et al. 1998, Slingsby and Waterman-Pearson 2002). Additional advantage from the preoperative administration of analgesic drugs is the reduction of the amount of anesthetics required for anesthesia induction and maintenance. The simultaneous administration of several analgesic classes or the combination of various analgesic treatment methods in order to attain pain relief is called multimodal analgesia. By this approach we gain advantage by combining the synergistic and additive effects that can be achieved when analgesics target different points along pain pathways (Lamont et al. 2000). Successful treatment of perioperative pain related with surgical excision of tumors has been associated with reduced incidence of metastasis in humans (Page et al. 1998). Although all tumors do not cause pain in humans it was found that 70% of oncologic patients, regardless of stage of disease and the chosen treatment, experience pain. In the cat, 32% of deaths attributed to neoplasms (Lascelles 2003). Thus, based on the similarities between human and animal perception of pain known as anthropomorphic approach (Anil et al. 2002) there is a clear need for pain alleviation in cats suffering from tumors.

Sometimes analgesia is necessary, especially in orthopedic disorders, where early mobilization of the operated limb is warranted for optimal recovery. Finally, one should not ignore the risk of injury to personnel or to the animal itself when it feels pain (Thurmon et al. 1996, Helleyer and Gaynor 1998).

TYPES OF FELINE PAIN AND THE IMPORTANCE OF PAIN CONTROL IN CATS

Analgesia in cats is essential to treat acute surgical, traumatic, perioperative and long-term pain. Long-term surgical pain may be related to tumors, degenerative joint disease, conditions of the oral cavity, non healing wounds and persistent postsurgical pain (Robertson and Lascelles 2010). Pain intensifies all neuroendocrine changes induced by trauma or surgery (Rollin 1997, Caroll et al. 1998). Energy requirements following surgical stress are increased resulting in a negative nitrogen balance and weight loss. Increased caloric intake should be implemented to combat these requirements. Pain control may significantly affect the cat's attitude and appetite leading to a positive energy balance, which is necessary for a rapid recovery and for a functional immune system (Thurmon et al. 1996, Rollin 1997, Caroll et al. 1998, Taylor and Robertson 2004).

Traditionally pain classification was based on two sets of descriptors. One category, based on duration, was acute pain that lasts for short time or chronic pain that lasts for at least 3 months and the other one was based on nervous involvement for pain transmission leading to the divisions of visceral, somatic and neuropathic pain (Crane 1987, Hardie and Kyles 1995, Hellyer et al. 2007). Acute pain is treated successfully with analgesic drugs, whereas chronic pain is often resistant to treatment (Hardie and Kyles 1995). Visceral pain transmission is accomplished from nociceptors located at viscera within the body cavities and somatic pain is transmitted through nociceptors located within locomotor system (Hardie and Kyles 1995). A more recent approach categorizes pain on the basis of the underlying mechanism as adaptive or maladaptive pain (Woolf 2004, Hellyer et al. 2007, Robertson and Lascelles 2010). Adaptive pain is an acute pain associated with inflammation following a surgical incision or traumatic incident. If this type of pain is not effectively addressed or if the duration of pain mismanagement is longer alterations in the central nervous system may increase pain sensitivity leading to maladaptive pain. Neuropathic pain is considered as an example for maladaptive pain (Robertson and Lascelles 2010). Pain conversion from adaptive to maladaptive may help practitioners understand the difficulties of pain management in certain patients (Hellyer et al. 2007). Under this classification scheme pain in the feline surgical patient may be individualized and different mechanisms of pain may coexist in the same patient requiring more than one treatment option (Hellyer et al. 2007, Robertson and Lascelles 2010). For example in a cat undergoing limb amputation inflammatory pain associated with the surgical incision and wound healing develops in the short term; if this pain is not appropriately controlled perioperatively using a multimodal approach it may switch to neuropathic long-term. Other components of maladaptive pain may include "wind-up pain" and allodynia (Hellyer et al. 2007). "Wind-up pain" is an increased sensitivity or worsening of an acute pain leading to changes in peripheral and central pain thresholds usually located at sites away from the primary source. Allodynia is termed the type of pain that results from a non painful stimulus. For example a cat that underwent a midline celiotomy incision may not tolerate even light stroking along its ventral abdomen long-term.

PAIN RECOGNITION

Pain recognition in the cat and in other animals it is not always easily performed even by experienced scientists (Hellyer and Gaynor 1998, Mathews 2000a, Lamont 2002). For this reason several pain scoring systems have been proposed, which could serve as guidelines for pain assessment. It is obvious that these systems have many shortcomings related with the subjectivity of the individual scientist to observe the animal in pain, the difficulty in interpretation of even objective clinical data and the different conditions occurring each time (Benson et al. 1991, Pibarot et al. 1997, McMillan 1998, Smith et al. 1999, Muir et al. 2004). These pain scoring systems are applied in dogs for several years but they are still in their infancy in cats. Slingsby et al. (2001) showed that the estimate of the intensity of postoperative pain in cats after orchiectomy, based on the reaction of animals when pressure was applied to the wound, was similar with the results from applying one of the systems for measuring pain (Visual Analogue Scale, VAS).

Behavioral changes are often a key criterion for pain recognition in the cat. In acute traumatic or postsurgical pain, cats react by restricting their movement and vocalization. After celiotomy, they prefer lying in sternal recumbency, they do not react to petting and they prefer to hide. They may intensely lick the painful area or exhibit lameness, when pain is located in an extremity (Dobromylskyj et al. 2000). Sometimes, especially young and otherwise healthy cats appear restless and become aggressive, as it sometimes happens after onychectomy (Hellyer and Gaynor 1998). Cat's purring should not mislead the clinician, as its existence does not exclude the presence of pain. Healing properties of feline purring are recently under investigation. Interestingly, it was shown that the frequency of the emitted sounds generated by purring are at similar frequency, which is used in some forms of physiotherapy including promotion of bone tissue growth and healing of fractures, (von Muggenthaler 2001). Many researchers correlated the presence of pain with parameters such as temperature, heart rate, respiratory rate or blood concentrations of cortisol, catecholamines and beta-endorphins. However, probably because affected by many other factors, these parameters have limited utility in the assessment of pain (Cambridge et al. 2000). For example, bandage application to a healthy limb although it does not cause a painful stimuli may serve as a potent stressful factor, able to result in an increase in urine cortisol by 200% (Taylor and Robertson 2004, Cloutier et al. 2005). Recently the use of methods for assessing pain in the cat in a laboratory setting were studied. One of these methods involves applying thermal stimuli to the skin of the animal (Dixon et al. 2002) and another method concerns measuring the force exerted by the limb when the animal walks on specific corridors [pressure-platform gait analysis] (Romans et al. 2004). The latter method has clinical implications regarding the pain derived from the extremities, such as pain following onychectomy, because these measurements reflect the way the cat uses its limbs.

Table 1. Postoperative pain assessment associated with various types of surgical procedures in cats

Severe pain	Moderate pain	Mild pain
Intercostal thoracotomy, medial sternotomy, orthopedic surgery, diaphragmatic herniorhaphy, limb amputation, total ear canal ablation, onychectomy	Midline celiotomy, flank laparotomy, ovariohysterectomy, dental surgery, tendonectomy, mastectomy, herniorhaphy, oncologic surgery	Castration, small mass removal, dental cleaning

Influence of Surgical Technique on Postoperative Pain

The extent of postoperative pain may be determined by combining the surgeon's experience concerning procedure's painfulness with the patient's behavioral and physiological response to surgery and by determining the individual patient response to pain relief strategies and treatments (Caroll 2007). Table 1 lists the most commonly performed surgical procedures in cats in terms of pain they elicit (Wright 2002, Williams et al. 2005, Hewson et al. 2006, Caroll 2007 Fossum).The majority of the pain levels summarized seems to be presumptive and may be affected by severity, inflammation or infection and surgical technique (Wright 2002). Irrespective of the pain estimate appeared in the table each animal should be evaluated in an individual basis and all cats that exhibit postoperative pain should receive analgesic therapy (Caroll 2007).

Ovariohysterectomy is a routine surgical procedure performed in cats to achieve population control. It is commonly carried out through a midline celiotomy or through a flank laparotomy technique that is mostly favored by British practitioners. A study performed in UK showed that analgesia provided for cats undergoing ovariohysterectomy was limited (Lachelles et al. 1999). Two clinical studies compared the severity of pain in cats undergoing ovariohysterectomy through a midline approach and flank approach reported that postoperative pain scores between the two techniques were similar (Grint et al. 2006, Burrow et al. 2006) but in one of these studies wound tenderness following the flank laparotomy was significantly more than that of midline approach (Grint et al. 2006). Differences in the postoperative pain between midline and flank approach may be attributed to less surgical trauma and less pain associated with limited visceral manipulations achieved through the midline incision compared to flank incision (Burrow et al. 2006). Flank laparotomy seems to be a more painful procedure in cats as the incision is significantly tenderer even when analgesia with non-steroidal anti-inflammatory agents is provided (Grint et al. 2006). As the previous study showed wound palpation seems to offer a reliable means for pain scoring (Grint et al. 2006). Minimally invasive surgery is also reported to provide a painless recovery in cats. Laparoscopic ovariohysterectomy seems to be a feasible sterilization technique in cats requiring no rescue analgesia within 24 hours after surgery (Kim et al. 2011).

Onychectomy is a commonly performed elective but controversial and painful procedure in cats aiming at avoiding undesirable scratching behavior (Gordon and Conzemius 2005). More than 14 million cats are declawed each year in the US (Patronek et al. 2001). Different surgical techniques were devised for cat declawing including scalpel blade excision, nail clippers (quilotine excision), radiofrequency unit excision and CO_2 laser excision. Cats during early recovery period from scalpel blade onychectomy may show reduced operated limb function because of pain (Romans et al. 2005). Recent evidence suggests that cats undergoing CO_2 laser excision seem to suffer less pain and showed improved limb function compared to cats undergoing scalpel blade excision (Holmbergh and Brisson 2006, Robinson et al. 2007).

Thoracic surgery in cats in not as commonly performed as in dogs but is considered a painful procedure. Postthoracotomy pain may lead to hypoventilation and prolonged recovery. Analgesia is necessary in all cats undergoing thoracotomy to provide pain control without adversely affecting respiration (Pavlidou et al. 2009). There is a lack of research regarding postoperative pain associated with thoracic surgery in cats in contrast to dogs but it is postulated that cats suffer postthoracotomy pain as much as dogs do. Type of surgical technique (open thoracotomy versus thoracoscopy), type of incision (median sternotomy versus intercostal thoracotomy) and technique used for rib or sternebrae approximation for thoracotomy closure may play a role in postoperative pain created (Pavlidou et al. 2009). Median sternotomy seems to create more pain that intercostal thoracotomy. Intercostal nerve trauma following muscle incision or rib retraction and nerve entrapment by sutures during rib approximation is mainly responsible for postthoracotomy pain (Pavlidou et al. 2009). Less nerve entrapment and consequently less pain creation may be achieved with blunt end needle passage in close proximity to the rib during intercostals thoracotomy closure (Rooney et al. 2004). Thoracostomy tube placement may also be a painful procedure since it traumatizes intercostal nerves (Pavlidou et al. 2009). Minimally invasive thoracoscopic surgery is considered less painful compared to open thoracotomy because of limited trauma created as a result of smaller incisions and no use of rib retraction (Pavlidou et al. 2009).

Analgesic Methods

Local Analgesia

Cats are not considered ideal candidates for the application of local or locoregional analgesia techniques. Due to their reactive character, the veterinarian often has to deeply sedate them in order to apply a technique of local anesthesia, thus reducing the benefit of the latter. Also, the small size of this species makes them more difficult to implement various techniques compared to other animal species and, finally, cats are sensitive to the toxic effects of local anesthetics (Rondenay and Cuvelliez 2002, Blaze and Glowaski 2004). The two most commonly used substances are lidocaine and boupivacaine. The first should not be administered at doses exceeding 6 mg / kg and its analgesic effect lasts about 1.5 hours, while boupivacaine's duration of action is approximately 4 hours and should not be administered at doses above 2 mg / kg. Both these drugs can be used for providing epidural analgesia. The tip of the spinal needle may sometimes inadvertently enter the subarachnoid space. The feline spinal cord extends farther caudally than it does in dogs allowing access to the epidural space via the lumbosacral foramen. If this happens, one must either reposition the needle or redefine the allocated amount of local anesthetic, eg, reducing it by half as the administration of the whole dose may result in spinal (subarachnoid) anesthesia. It was also found that the concomitant administration of epidural boupivacaine and morphine significantly prolongs analgesia (Troncy et al. 2002). It is reported that skin application of an ointment containing lidocaine and prilocaine provides analgesia sufficient to facilitate the placement of intravenous catheters, preventing the active substances to be absorbed into the systemic circulation, or cause the occurrence of local adverse reactions (Gibbon et al. 2003). Clinical experience, however, shows that the induced analgesia is not very satisfactory in the cat. Furthermore, the use of local anesthetics for wound irrigation did not provide postoperative analgesia in cats undergoing onychectomy (Winkler et al. 1997). Wound infiltration with a local anesthetic was reported to provide effective analgesia in dogs (Savvas et al. 2008) but was not effective in cats undergoing midline celiotomy (Tobias et al. 2006). Wound infiltration with a continuous administration of a local anesthetic was not studied in cats as in dogs (Wolfe et al. 2006)

Systemic Analgesia

Non steroid anti-inflammatory drugs are among the most commonly prescribed veterinary formulations. These drugs act by inhibiting the prostaglandin production from arachidonic acid by the cyclooxygenase (COX) enzymes. The COX enzymes exists in three isoforms: COX-1 are constantly produced and are responsible for the physiological actions of prostaglandins, COX-2 and the recently discovered COX-3, induced by inflammatory tissues are responsible for causing inflammation (Schwab et al. .2003a, 2003 b). In recent years drugs with greater specificity to the inhibition of COX-2 compared with COX-1 (carprofen, meloxicam) or even with action only against COX-2 (robenacoxib) were produced. However, side effects still occur after administration of even more specific inhibitors (Mathews 2000b). The cat shows peculiarities in the metabolism of various substances as NSAIDs. In particular the metabolic pathway of glucuronic acid is practically inactive in the cat, so many drugs either metabolized very slowly (acetylsalicylic acid), or are particularly toxic, like these that in their molecules containe a phenolic ring, e.g. paracetamol (Court and Greenblatt 1997a, 1997b). It was also shown that the half-life of carprofen in the cat is approximately 20 hours, twice than that of the dog and similar to acetylsalicylic acid (Taylor et al. 1996, Parton et al. 2000). The risk of toxicity associated with NSAIDs can be prevented by appropriate dosing and intervals between administered doses. Continuous research on the metabolism of NSAIDs in cats, their relative low cost, increased duration of action, efficacy in controlling pain, the lack of restrictions on marketing as opposed to opioids and co-administration with opioids, can make NSAIDs an attractive and perhaps the first choice for pharmaceutical pain control in cats, especially in the postoperative period (Lascelles et al. 1995, Balmer et al. 1998, Slingsby and Waterman - Pearson 1998, 2000, 2002). However, in disease situations such as dehydration, kidney or liver failure, hypotension, heart failure, ascites, thrombocytopenia, bleeding, diseases of the digestive tract, shock, and perhaps bronchial disease their use is contraindicated (Carroll and Simonson 2005).

Opioids act at receptors in the central nervous system thereby causing "central" analgesia. These receptors are divided into several categories (μ, k, δ, etc.). Stimulation of mu-receptors results in dose-dependent analgesia, particularly effective in physical pain, while stimulation of k-receptors may lead to milder analgesia, effective in visceral pain (Pascoe 2000). Substances, which act on the k-receptors, exhibit the ceiling phenomenon, namely increasing the quantity allocated over a certain threshold, seem not to cause further increase in the intensity of their action (analgesic, sedative, respiratory or other), but only extend their action. Since opioids act on receptors located in the spinal cord are effective if administered in the epidural space. Recently the effect of opioids in peripheral receptors is investigated.

Antagonists of NMDA-receptors in the spinal cord such as ketamine apart from achieving analgesia, are believed that their administration may prevent and even reverse central sensitization (Gaynor 2002). Ketamine was shown that at a small analgesic dose may cause a mild analgesic effect (Waterman-Pearson 1999). Prolongation of its action postsurgery can be achieved by a continuous drip (Gaynor 2002).

References

Anil, SS, Anil, L and Deen, J (2002). Challenges of pain assessment in domestic animals. *Journal of the American Veterinary Medical Association*, 220: 313-319.

Balmer, TV, Irvine, D, Jones, RS, Roberts, MJ, Slingsby, L, Taylor, PM and Waterman AE, Waters C (1998). Comparison of carprofen and pethidine as postoperative analgesics in the cat. *Journal of Small Animal Practice,* 39: 158-64.

Benson, GJ, Wheaton, LG, Thurmon, JC, Tranquilli, WJ, Olson, WA and Davis, CA (1991). Postoperative catecholamine response to onychectomy in isoflurane-anesthetized cats. Effects of analgesics. *Veterinary Surgery*, 20: 222-225.

Blaze, CA and Glowaski, MM (2004). Individual drugs. In: Blaze CA and Glowaski MM (eds), Veterinary Anesthesia. Drug Quick Reference (pp 1-100). St. Louis: Elsevier, Saunders.

Burrow R, Wawra E, Pinchbeck. G, Senior. M and Dugdale, A (2006). Prospective evaluation of postoperative pain in cats undergoing ovariohysterectomy by a midline or flank approach. *The Veterinary Record*, 158: 657-660.

Cambridge, AJ, Tobias, KM, Newberry, RC and Sarkar DK (2000). Subjective and objective measurements of postoperative pain in cats. *Journal of the American Veterinary Medical Association*, 217: 685-690.

Caroll, GL (2007). Perioperative multimodal analgesic therapy. In: Fossum TW (ed), Small Animal Surgery (3rd edn, pp 130-145). Saint Louis: Mosby, Elsevier.

Carrol, GL, Howe, LB, Slater, MR, Haughn, L, Martinez, EA, Hartsfield, SM and Matthews NS (1998). Evaluation of analgesia provided by postoperarive admistration of butorphanol to cats undergoing onychectomy. *Journal of the American Veterinary Medical Association*, 213: 246-250.

Carrol, GL and Simonson, SM (2005). Recent developments in nonsteroidal antiinlammatory drugs in cats. *Journal of the American Animal Hospital Association*, 41: 347-354.

Cloutier, S, Newberry, RC, Cambridge, AJ and Tobias KM (2005). Behavioural signs of postoperative pain in cats following onychectomy or tenectomy surgery. *Applied Animal Behaviour Science*, 92: 325-335.

Court, MH and Greenblatt DJ (1997α). Biochemical basis for deficient paracetamol glucuronidation in cats: an interspecies comparison of enzyme constraint in liver microsomes. *Journal of Pharmacy and Pharmacology*, 49: 446-449.

Court, MH and Greenblatt, DJ (1997β). Molecular basis for deficient acetaminophen glucuronidation in cats. An interspecies comparison of enzyme kinetics in liver microsomes. *Biochemical Pharmacology*, 53: 1041-1047.

Crane, SW (1987). Perioperative analgesia: a surgeon's perpective. *Journal of the American Veterinary Medical Association*, 191: 1254-1257.

Dixon, MJ, Robertson, SA and Taylor, PM (2002). A thermal threshold testing device for evaluation of analgesics in cats. *Research in Veterinary Science*, 72: 205-210.

Dobromylskyj, P, Flecknell, PA, Lascelles, BD, Livingston, A, Taylor, P and Waterman-Pearson A (2000). Pain assessment. In: Flecknell PA and Waterman-Pearson A (eds), Pain Management in Animals (pp 53-80). London: WB Saunders.

Gaynor, JS (2002). Other drugs used to treat pain. In: Gaynor, JS and Muir, WW (eds), Handbook of Veterinary Pain Management (1st edn, pp 251-260). St Louis: Mosby.

Gibbon, KJ, Cyborski, JM, Guzinski, MV, Viviano, KR and Trepanier, LA (2003). Evaluation of adverse effects of EMLA (lidocaine/prilocaine) cream for the placement of jugular catheters in healthy cats. *Journal of Veterinary Pharmacology and Therapeutics,* 26: 439-441.

Gordon, WJ and Conzemius, MG (2005). Analgesia after onychectomy in cats. *Veterinary Medicine*, 100: 46-54.

Grint, NJ, Murison, PJ, Coe, RJ and Waterman - Pearson AE (2006). Assessment of the influence of the surgical technique on postoperative pain and wound tenderness in cats following ovariohysterectomy. *Journal of Feline Medicine and Surgery*, 8: 15-21.

Hardie, EM and Kyles, AE (1995). Pharmacological management of pain and infection in the surgical patient. *Veterinary Clinics of North America: Small Animal Practice*, 25: 77-96.

Hellyer, PW and Fails, AD (2003). Pain management for the surgical patient. In: Slatter, D (ed), Textbook of Small Animal Surgery (1st edn, pp 2503-2515). Philadelphia: Saunders.

Hellyer. PW and Gaynor. JS (1998). Acute postsurgical pain in dogs and cats. Compendium on Continuing Education for the Practicing Veterinarian, 20: 140-153.

Hellyer, P, Rodan, I, Brunt, J, Downing, R, Hagerdorn, JE and Robertson, SA (2007). AAHA/AAFP pain management guidelines for dogs and cats. *Journal of Feline Medicine and Surgery*, 9: 466-480.

Hewson, CJ, Dohoo, IR and Lemke, KA (2006). Perioperative use of analgesics in dogs and cats by Canadian veterinarians. *Canadian Veterinary Journal*, 47: 352-359.

Holmberg, DL and Brisson, BA (2008). A prospective comparison of postoperative morbidity associated with the use of scalpel blades and lasers for onychectomy in cats. *Canadian Veterinary Journal*, 47: 162-163.

Hugonnard, M, Leblond, A, Keroack, S, Cadore, JL and Troncy, E (2004). Attitudes and concerns of French veterinarians towards pain and analgesia in dogs and cats. *Veterinary Anaesthesia and Analgesia*, 31: 154-163.

Joubert, KE (2006). Anaesthesia and analgesia for dogs and cats in South Africa undergoing sterilization and with osteoarthritis-an update from 2000. *Journal of the South African Veterinary Association*, 77: 224-228.

Kim, YK, Lee, SY, Park, SJ, Lee, SS, Lee, HC, Lee, HJ and Yeon, SC (2011). Feasibility of single –portal access laparoscopic ovariectomy in 17 cats. *The Veterinary Record*, 169: 179.

Lamont, LA (2002). Feline perioperative pain management. *Veterinary Clinics of North America Small Animal Practice*, 32: 747-763.

Lamont, LA, Tranquilli, WJ and Mathews, KA (2000). Adjunctive analgesic therapy. *Veterinary Clinics of North America Small Animal Practice*, 30: 805-813.

Lascelles, BD, Capner, CA and Waterman-Pearson, AE (1999). Current British veterinary attitudes to perioperative analgesia for cats and small mammals. *The Veterinary Record*, 145: 601-604.

Lascelles, BD, Cripps, P, Mirchandani, S and Waterman, AE (1995). Carprofen as an analgesic for postoperative pain in cats: dose titration and assessment of efficacy in comparison to pethidine hydrochloride. *Journal of Small Animal Practice*, 36: 535-541.

Lascelles, BD, Robertson, SA, Taylor, PM and Hauptman, JG (2003). Thermal antinociceptive pharmacodynamics of 0.1 mg/kg hydromorphone administered intramuscularly in cats, and effect of concurrent butorphanol administration. *Veterinary Anaesthesia and Analgesia*, 30: 108-109.

Livingston, A, Champers, P (2000). The physiology of pain. In: In: Flecknell PA and Waterman-Pearson A (eds), Pain Management in Animals (pp 9-20). London: WB Saunders.

Maps of the world. Top ten countries with most pet cat population (2006). http://www.mapsofworld.com/world-top-ten/countries-with-most-pet-cat-population.html

Mathews, KA (2000a). Pain assessment and general approach to management. *Veterinary Clinics of North America Small Animal Practice*, 30: 729-756.

Mathews, KA (2000b). Nonsteroidal anti-inflammatory analgesics. Indications and contraindications for pain management in dogs and cats. *Veterinary Clinics of North America Small Animal Practice*, 30: 783-804.

McMillan, FD (1998). Comfort as the primary goal in veterinary medical practice. *Journal of the American Veterinary Medical Association*, 212: 1370-1374.

Muir, WW III (2002). Physiology and pathophysiology of pain. In: Gaynor, JS and Muir, WW (eds), Handbook of Veterinary Pain Management (1st edn, pp 13-45). St Louis: Mosby.

Muir,WW III, Wiese, AJ and Wittum, TE (2004). Prevalence and characteristics of pain in dogs and cats examined as outpatients at a veterinary teaching hospital. *Journal of the American Veterinary Medical Association*, 224: 1459-1463.

Page, GG, McDonald, JS and Ben-Eliyahu S (1998). Pre-operative versus postoperative administration of morphine: impact on the neuroendocrine, behavioural, and metastatic-enhancing effects of surgery. *British Journal of Anaesthesia*, 81: 216-223.

Parton, K, Balmer, TV, Boyle, J, Whittem, T and MacHon, R (2000). The pharmacokinetics and effects of intravenously administered carprofen and salicylate on gastrointestinal mucosa and selected biochemical measurements in healthy cats. *Journal of Veterinary Pharmacology and Therapeutics*, 23: 73-79.

Pascoe, PJ (2000). Opioid analgesics. *Veterinary Clinics of North America Small Animal Practice*, 30: 757-773.

Patronek, CJ (2001). Assessment of claims of short –and long-term complications associated with onychectomy in cats. *Journal of the American Veterinary Medical Association*, 219: 932-937.

Pavlidou, K, Papazoglou, LG, Savvas, I and Kazakos, G (2009). Analgesia for small animal thoracic surgery. *Compendium: Continuing Educating for Veterinarians*, 31: 432-436.

Pibarot, P, Dupuis, J, Grisneaux, E, Cuvelliez, S, Plante, J, Beauregard, G, Bonneau, NH, Bouffard, J and Blais, D (1997). Comparison of ketoprofen, oxymorphone hydrochloride, and butorphanol in the treatment of postoperative pain in dogs. *Journal of the American Veterinary Medical Association*, 211: 438-444.

Raekallio, M, Heinonen, KM, Kuussaari, J and Vainio, O (2003). Pain alleviation in anaimls:attitudes and practices of Finish veterinarians. *The Veterinary Journal*, 165: 131-135.

Raptopoulos, D, Savvas, I and Papazoglou L (2002). Pain Management: From Theory to Practice. Proceedings, Perioperative Analgesia, Thessaloniki, pp A1-A13. (In Greek with English abstract).

Robinson, DA; Romans, CW; Gordon-Evans, WJ and Conzemius, MG (2007). Evaluation of short-term limb function following unilateral carbon dioxide laser or scalpel onychectomy in cats. *Journal of the American Veterinary Medical Association*, 230: 353-358.

Robertson, S and Lascelles, D (2010). Long-term pain in cats. *Journal of Feline Medicine and Surgery*, 12: 188-199.

Rollin, BE (1997). Pain and ideology in human and veterinary medicine. *Seminars in Veterinary Medicine and Surgery*, 12: 56-60.

Romans, CW, Conzemius, MG, Horstman, CL, Gordon, WJ and Evans, RB (2004). Use of pressure platform gait analysis in cats with and without bilateral onychectomy. *American Journal of Veterinary Research,* 65: 1276-1278.

Romans, CW, Gordon, WJ, Robinson, DA, Evans, R and Conzemius, MG (2005). Effect of postoperative analgesic protocol on limb function following onychectomy in cats. *Journal of the American Veterinary Medical Association*, 227: 89-93.

Rondenay, Y and Cuvelliez, S (2002). Feline Anesthesia. In: Greene, SA (ed), Veterinary Anesthesia and Pain Management Secrets (pp239-244). Philadelphia: Hanley and Belfus Inc.

Rooney, MB; Mehl, M and Monnet, E (2004). Intercostal thoracotomy closure: transcostal sutures as a less painful alternative to circumcostal suture placement. *Veterinary Surgery*, 33: 209-213.

Savvas, I, Papazoglou, LG, Kazakos, G, Anagnostou, T, Tsioli, V and Raptopoulos D (2008). Incisional block with bupivacaine for analgesia following celiotomy in dogs. *Journal of the American Animal Hospital Association*, 44: 60-66.

Schwab, JM, Beiter, T, Linder, JU, Laufer, S, Schulz, JE, Meyermann, R and Schluesener HJ (2003a). COX-3–a virtual pain target in humans? *The FASEB Journal*, 17: 2174-2175.

Schwab, JM, Schluesener, HJ and Laufer S (2003b). COX-3: just another COX or the solitary elusive target of paracetamol? *Lancet*, 361: 981-982.

Slingsby, LS, Lane, EC, Mears, ER, Shanson, MC and Waterman-Pearson AE (1998). Postoperative pain after ovariohysterectomy in the cat: a comparison of two anaesthetic regimens. *The Veterinary Record*, 143: 589-590.

Slingsby, LS and Waterman-Pearson, AE (1998). Comparison of pethidine, buprenorphine and ketoprofen for postoperative analgesia after ovariohysterectomy in the cat. *The Veterinary Record*, 143: 185-189.

Slingsby, LS and Waterman-Pearson, AE (2000). Postoperative analgesia in the cat after ovariohysterectomy by use of carprofen, ketoprofen, meloxicam or tolfenamic acid. *Journal of Small Animal Practice*, 41: 447-450.

Slingsby, LS, Jones, A and Waterman-Pearson, AE (2001). Use of a new finger-mounted devise to compare mechanical nociceptive thresholds in cats given pethidine or no medication after castration. *Research in Veterinary Science*, 70: 243-246.

Slingsby LS and Waterman-Pearson, AE (2002). Comparison between meloxicam and carprofen for postoperative analgesia after feline ovariohysterectomy. *Journal of Small Animal Practice*, 43: 286-289.

Smith, JD, Allen, SW and Quandt, JE (1999). Changes in cortisol concentration in response to stress and postoperative pain in client-owned cats and correlation with objective clinical variables. *American Journal of Veterinary Research*, 60: 432-436.

Taylor, PM, Delatour, P, Landoni, FM, Deal, C, Pickett, C, Shojaee Aliabadi, F, Foot, R and Lees P (1996). Pharmacodynamics and enantioselective pharmacokinetics of carprofen in the cat. *Research in Veterinary Science*, 60: 144-151.

Taylor, PM and Robertson, SA (2004). Pain management in cats-past, present and future. Part 1. The cat is unique. *Journal of Feline Medicine and Surgery*, 6: 313-320.

Thurmon JC, Tranquilli WJ and Benson, GJ (1996). Perioperative pain and distress. In: Thurmon JC, Tranquilli WJ and Benson, GJ (eds), Lumb and Jones' Veterinary Anesthesia (3rd edn, pp 40-62). Baltimore: Williams and Wilkins.

Tobias, KM, Harvey, PC Byarlay, MJ (2006). A comparison of four methods of analgesia in cats following ovariohysterectomy. *Veterinary Anaesthesia and Analgesia*, 33: 390–398.

Troncy, E, Junot, S, Keroack, S, Sammut, V, Pibarot, P, Genevois, JP and Cuvelliez S (2002). Results of preemptive epidural administration of morphine with or without bupivacaine in dogs and cats undergoing surgery: 265 cases (1997-1999). *Journal of the American Veterinary Medical Association*, 221: 666-672.

von Muggenthaler (2001). The felid purr: A healing mechanism? *Journal of the Acoustical Society of America*, 110: 2666.

Waterman-Pearson, AE (1999). Analgesia. In: Seymour, C and Gleed, R (eds), Manual of Small Animal Anaesthesia and Analgesia (2nd edn, pp 59-70). Cheltenham: BSAVA.

Williams, VM; Lascelles, BDX and Robson, MC (2003). Current attitudes, and use of, peri-operative analgesia in dogs and cats by veterinarians in New Zealand. *New Zealand Veterinary Journal*, 53: 193-202.

Winkler, KP, Greenfield, CLand Benson GJ (1997). The effect of wound irrigation with bupivacaine on postoperative analgesia of the feline onychectomy patient. *Journal of the American Animal Hospital Association*, 33: 346-352.

Wolfe, TM, Bateman, SW, Cole, LK and Smeak, DD (2006) Evaluation of a local anesthetic delivery system for the postoperative analgesic management of canine total ear canal ablation - a randomized, controlled, double-blinded study._*Veterinary Anaesthesia and Analgesia*, 33: 328-339.

Woolf, CJ (2004). Pain: moving from symptom control toward mechanism-specific pharmacologic management. *Annals of Internal Medicine*, 140: 441-451.

Wright, BD (2002). Clinical pain management techniques in cats. *Clinical Techniques in Small Animal Practice*, 17: 151-157.

In: Cats
Editors: T. Zakrevsky and V. Lachinov

ISBN 978-1-62081-043-9

Chapter IV

Streptococcus Equi Subspecies *Zooepidemicus*: An Emerging Pathogen of Cats

Ann P. Britton[1] and Jennifer L. Davies[2]
[1]Animal Health Centre, British Columbia Ministry of Agriculture, Abbotsford, BC, Canada
[2]Diagnostic Services Unit, Faculty of Veterinary Medicine, University of Calgary, Calgary, AB, Canada

Abstract

Streptococcal disease is a common occurrence in animals. Most streptococcal species are host specific causing sporadic expression of disease in the host. *Streptococcus equi* subspecies *zooepidemicus* (SEZ) is a mucosal commensal in a wide range of animals, most significantly in horses, but is not a recognized commensal of the dog or the cat. Since the early 1980's, there have been increasing reports of SEZ linked to outbreaks of severe fatal respiratory disease in dogs maintained in shelters, research facilities, and kennels. The bacterium has been denoted an emerging pathogen of dogs kept in intensively housed environments. In 2010, a similar clinical presentation was reported in cats maintained in shelters. Similar to intensively housed dogs, sheltered cats exhibited severe fatal pneumonia, but were also reported to contract rhinitis and in some cases meningitis. SEZ is now considered an emerging pathogen of cats. In this chapter we discuss SEZ and expression of disease in the cat.

Introduction

Streptococcal bacteria are Gram positive, nonmotile, facultatively anaerobic cocci many of which can cause a wide range of pyogenic infections in animals and man [1]. Streptococcal bacteria are associated with sporadic disease in individuals as well as outbreaks of disease in both animals and humans living in intensive housing situations. Numerous classification systems have been developed for streptococci of which hemolytic activity and the Lancefield serogrouping of cell wall carbohydrates are most commonly utilized [1]. Hemolytic activity is divided into three classes: non-hemolytic, alpha-hemolytic and beta-hemolytic, of which beta-hemolytic strains are considered most pathogenic. The Lancefield method subdivides streptococcal species into 7 groups denoted A,B,C,E,G,L and M [1].

Streptococcal Disease

Streptococcal disease is uncommon in cats [2]. Beta-hemolytic, Group G streptococcal infection, usually *S. canis*, is the most commonly reported feline streptococcal pathogen. *S. canis* is known to be a commensal of the feline skin, oropharynx, and urogenital tract suggesting that disease expression in cats is likely opportunistic. Two studies of oropharyngeal and rectal commensalism in normal cats revealed 6% and 19% had oropharyngeal carriage and 13% and 30% had rectal carriage of *S. canis* [3, 4] indicating that carriage is not present in the majority of cats and that prevalence of carriage can vary. Two retrospective studies of infectious agents associated with chronic rhinitis and rhinosinusitis in cats and normal control cats failed to recover any beta-hemolytic streptococci [5, 6] suggesting that low prevalence of commensalism may also contribute to low prevalence of feline streptococcal disease.

S. canis is associated sporadically with feline neonatal sepsis, urogenital infections, skin wounds, abscesses [7], pyothorax and pleuritis [2]. Other disease conditions such as arthritis [8], lymphadentitis [9], necrotizing fasciitis [10, 11], rhinosinusitis and meningitis [11] are most often observed in outbreak form in cats maintained in intensively housed conditions such as cat shelters and cat colonies. In two outbreaks with high mortality, suppurative and necrotizing sinusitis and rhinitis also lead to fatal meningitis in many affected cats [11]. It is unknown what virulence factors may be involved with invasive streptococcal disease, but the recovery of *S. canis* biotype 1 from all affected cats in these two outbreaks suggests that there may be strain specific virulence factors [11].

In 2010, pneumonia, rhinitis and meningitis associated with a beta-hemolytic, group C streptococcal infection, *Streptococcus equi* subspecies *zooepidemicus* (SEZ), was reported in cats housed in Israeli and Canadian shelters for the first time [12, 13]. This followed in the footsteps of increasing numbers of outbreaks of severe fatal respiratory infections with SEZ in dogs maintained in shelters [14, 15, 16], research facilities [17, 18] and kennels [19] indicative of emerging pathogen status of this bacterium for dogs maintained in intensively housed environments.

Streptococcus Equi Subspecies Zooepidemicus: Overview

SEZ is a mucosal commensal with a wide host range [20] and occasionally strains can invade host tissue or cross species barriers [21]. The bacterium is the most commonly isolated opportunistic pathogen of horses and is associated with pneumonia, pleuropneumonia, endometritis, arthritis, neonatal septicemia, abortion, mastitis [22] and meningoencephalitis

[23]. Although SEZ is reported to be a commensal in a wide range of animals, it is not a recognized commensal of the dog [4, 24] or the cat [4]. How dogs and cats in shelter environments come to be infected with an equine/farm animal commensal is unclear as no equine or other farm animal exposure could be demonstrated in the reported outbreaks. However, SEZ was isolated from cats with mild respiratory disease during the Israeli shelter outbreak and from oral swabs of dogs housed adjacent to fatally infected dogs in the shelter outbreak reported by Pesavento *et al* in 2008 indicating that once exposure is achieved, SEZ may result in non fatal infection and possibly carrier status in some cats and dogs [12, 15].

Individual pet dogs are rarely affected by fatal SEZ pneumonia [25, 26]. Residence in shelters, research laboratories, and other facilities where close confinement of animals occurs is the major risk factor for development of severe SEZ associated disease in dogs [14, 15]. Why close confinement may predispose to SEZ disease is unclear. Stress arising from transporting animals to new environments, mixing unfamiliar animals from various origins and preexistent viral or bacterial infection may predispose to disease expression in the presence of an asymptomatic carrier [27]. Transportation of horses over long distances can result in severe hemorrhagic to necrotizing bronchopneumonia associated with aerogenous spread of SEZ from the tonsils to the lower respiratory tract [28]. While unaffected horses reveal mixed bacteria in tonsils following transport, pneumonic horses have fewer bacterial species with a much heavier growth of SEZ in their tonsils suggesting that transportation stress may favour tonsillar proliferation and spread of SEZ to the lower respiratory tract in certain individuals [29].

Underlying viral disease can often play a role in expression of bacterial disease. Experimental reproduction of mild non-fatal SEZ pneumonia has been demonstrated in dogs with intranasal installation of SEZ and canine influenza virus whereas intranasal installation of SEZ alone failed to cause significant disease [30]. While influenza infection potentiated mild SEZ disease in the canine study, it did not replicate the severe hemorrhagic pneumonia observed in shelters [30]. This is compatible with observations that co infection with other respiratory pathogens was unrelated to expression of fatal SEZ respiratory disease in dogs in natural outbreaks [14, 15, 17] suggesting that the introduction of either a high challenge or a virulent clone of SEZ into an environment densely populated with susceptible hosts [15] is most likely responsible for the severe disease expression. Unlike intranasal inoculation, experimental intratracheal installation of SEZ in dogs will reproduce severe fatal hemorrhagic pneumonia [17] further suggesting that introduction of a high challenge to the lower respiratory tract likely plays an important role in pathogenesis.

VIRULENCE

Currently there is insufficient data to determine the basis for high virulence of SEZ in dogs [27] and cats. Interestingly, multilocus sequence typing (MLST) found that several isolates of SEZ causing outbreaks of fatal hemorrhagic pneumonia in dogs from geographically remote locations were genetically related. This finding suggests that certain strains of SEZ are more adept at causing disease in dogs [21]. To date, this type of molecular analysis of feline isolates has not been reported and it remains unknown as to whether particular strains of SEZ are associated with disease in cats.

In general, the virulence of pathogenic streptococci is based on surface structures that either prevent phagocytosis or are involved in adhesion; the best-understood factors are the hyaluronic acid capsule and the antiphagocytic M proteins. Many pathogenic streptococci also have the ability to bind components of the host's plasma, and organisms coated with these components may be able to escape detection of the host's immune system. Specific virulence factors associated with SEZ include capsular hyaluronic acid, streptokinase, proteases, streptolysin S, peptidoglycan, fibronectin binding protein, IgG protein and SzP protein [20].

The cell surface protein, SzP, is an M-like protein and is important in the pathogenesis of infection in equids. The SzP protein binds equine fibrinogen which aids in preventing phagocytosis and limits deposition of C3 on the cell surface [31]. In addition, studies have shown that variation in the SzP protein is a useful measure of diversity among strains of SEZ [32]. However, while SzP is an important virulence factor in equids there does not appear to be a direct relationship between variations in SzP protein and particular disease manifestations in horses [31] nor is SzP phenotype an important determinant of invasiveness or epizootic capabilities [32]. To the best of the authors' knowledge, variations in SzP proteins of SEZ isolates from cats and dogs have not been examined.

Gene gain via the horizontal acquisition of mobile genetic elements is a key factor in the emergence of pathogenic strains of streptococci. In humans, the emergence of strains of *S. pyogenes* causing streptococcal toxic shock syndrome (STSS) is associated with the acquisition of genes encoding for superantigens [33]. *S. pyogenes* produces several pyogenic exotoxins which act as superantigens and are associated with increased virulence of this organism. Unlike conventional antigens, superantigens have high immunomodulating capacity [20]. Superantigens activate a large portion of T-lymphocytes leading to exuberant cytokine production and activation of neutrophils and vasoactive factors. These factors act on endothelium to increase permeability, cause vasodilation and promote coagulation [27].

In dogs with SEZ pneumonia, the rapid onset of disease and fast deterioration is similar to STSS in man [27]. Recently three novel superantigens, SzeP, SzeN and SzeF, have been identified from a strain of SEZ isolated from a case of acute fatal hemorrhagic pneumonia in a dog from the UK. This suggests that horizontal transfer of genes encoding for superantigens within the diverse SEZ population could result in the emergence of virulent strains. However, this study was unable to link the presence of superantigen-encoding genes in the broader SEZ population to cases of acute fatal hemorrhagic pneumonia of dogs. Therefore, currently there is insufficient data to determine if superantigens or other bacterial toxins are involved in the pathogenesis of SEZ infections in dogs [27]. To date, the genome of feline SEZ isolates has not been examined for the presence of superantigens, and it is unknown whether these factors may be associated with the emergence and severe disease manifestations of SEZ now being observed in cats.

Pneumonia

The majority of feline cases of SEZ disease published to date occurred in one cat shelter with severe pneumonia presenting in outbreak form [12]. Thirty-nine cats in this outbreak exhibited clinical signs characterized by nasal discharge, cough and/or dyspnea prior to death [12]. Necropsy findings described severe acute diffuse suppurative to necrosuppurative bronchopneumonia with intralesional coccoid bacteria similar to that described for dogs and transported horses. SEZ was isolated from all affected lungs, usually in pure culture, but was not isolated from 29 other cats which died without respiratory disease during the outbreak [12]

Infectious pneumonia is uncommon in cats. Retrospective studies examining causes of pneumonia in cats reveal that pneumonia was diagnosed as the cause of death in only 3.5% of cases (110/31,323) over the period 1991 to 2000 and of those cases only 35% (39/110) were infectious [34]. Of the bacterial causes of pneumonia, *S. canis* and *Pasteurella multocida* were isolated most commonly, each representing 29% of the cases [34]. One additional case yielded a non speciable beta-hemolytic streptococcus [34]. In contrast, a second study looking at cats with non-fatal pneumonia over the period 1995 to 2000 found no streptococcal involvement [35]. Recovery of beta-hemolytic streptococci, whether *S. canis* or SEZ, only from fatal bacterial pneumonia in cats suggests that these bacteria have significant virulence.

In the retrospective study of fatal pneumonia in cats, severity of clinical signs correlated closely with severity of pulmonary pathology, but did not correlate significantly with the infectious agents recovered [34]. Fewer than half of the cats with infectious pneumonia presented with clinical signs indicative of respiratory disease (dyspnea, tachypnea, cough and/or nasal discharge) and the other half had no clinical signs whatsoever [34]. This exceedingly variable clinical expression of lower respiratory tract disease is typical of cats and can make detection and diagnosis challenging [36]. In contrast, the shelter outbreak of SEZ disease describes nasal discharge, cough and/or dyspnea prior to death in 100% of cats with pneumonia [12] reflecting
the severity of the pneumonia. Thus when respiratory tract infection is either detected or suspected in cats, inclusion of appropriate antimicrobial therapy aimed at streptococcal disease would be indicated.

Rhinitis and Meningitis

Rhinitis and meningitis have been reported in sporadic cases of SEZ in shelter cats [12, 13]. Although the two cats reported by Britton and Davies had short clinical histories prior to death due to meningitis, the rhinitis in both cats was characterized by increased numbers of plasma cells in the nasal mucosa indicative of chronicity. Myeloid hyperplasia was noted in the bone marrow of both cats indicative of chronic demand for leucocytes, which was concluded to have been associated with the rhinitis as no other chronic source of infection was found in the cats [13]. This suggests that SEZ disease in cats may commence with rhinitis, which if not cleared by appropriate antibiotic treatment, may ultimately spread to the brain and/or the lung.

Chronic rhinitis associated with SEZ infection was recently reported in two pet dogs resident on horse farms, one of which also exhibited chronic pneumonia [37, 38]. A third dog on one of the horse farms was positive on oropharyngeal culture for SEZ but exhibited no clinical signs, indicative of the potential for asymptomatic carrier status [37]. Interestingly the asymptomatic dog was colonized by SEZ strain ST173, the same virulent strain isolated from an outbreak of hemorrhagic pneumonia in shelter dogs [39]. All dogs were responsive to antibiotic treatment including the dog with chronic pneumonia [37, 38]. Emergence of sporadic non-fatal disease and asymptomatic carrier status in dogs on horse farms suggests that SEZ infection may indeed initially arise from exposure to horses but that, in addition to the strain of SEZ, other factors play a role in the severity of disease expression.

Streptococcal meningitis is described in both humans and animals and usually occurs as a result of bacteremia and breaching of the blood brain barrier [40]. However, experimental spread of *Streptococcus pneumoniae* from rhinitis along olfactory neurons to the brain has been achieved in mice [41]. In the two outbreaks of *S. canis* meningitis in shelter cats and in one cat with SEZ rhinitis and meningitis, inflammatory damage to the nasal bones and cribriform plate was present and the authors concluded that the meningitis, in the absence of pneumonia, most likely occurred due to olfactory spread of bacteria through the cribriform plate to the brain [11, 13]. Interestingly, lysis of nasal turbinate was also identified in a dog with chronic rhinitis associated with SEZ infection but no meningitis was identified [38]. SEZ meningitis has also been reported in a pet cat with otitis media/interna suggesting another possible route of infection to the brain via the auditory pathway [42].

SEZ is a zoonotic bacterium which most often presents as meningitis in humans. Although there is widespread SEZ carriage in horses and widespread exposure of humans to horses, zoonotic infection in man is uncommon [27]. Of 19 human SEZ cases reviewed in the literature, 14 had meningitis of which 9 were exposed to horses and 5 ate infected unpasteurized milk or cheese [43]. One recent case of upper respiratory infection occurred in a human exposed to a dog with chronic rhinitis and pneumonia subsequent to the dog sneezing in the human's face [39]. Strain ST178 of SEZ was identified in both the dog and the human, distinct from the ST-173 strain identified in outbreaks of shelter dogs, again suggesting that degree of virulence may be strain related [39]. In a retrospective study of strains of SEZ associated with disease from multiple countries, it was found that isolates from 8 outbreaks of acute fatal hemorrhagic pneumonia in British and American dogs were genetically related, but not identical, suggesting that certain subtypes of SEZ may be more virulent for dogs [21]. Five of these canine strains were also recovered from equine disease [21] compatible with a study which found it unlikely that a single virulent strain of SEZ is responsible for equine pneumonia [32]. Interestingly, cats were housed at two of the dog shelters which experienced outbreaks of fatal hemorrhagic SEZ pneumonia [15, 16]. All cats were clinically unaffected in one outbreak [16], while one cat died in the second outbreak carrying a different strain of SEZ to that affecting the dogs [15]. This suggests that different strains of SEZ may be involved in feline expression of the disease or that other factors may precipitate disease in the cat.

Conclusion

SEZ is a bacterium which has recently been designated an emerging pathogen of cats causing a disease spectrum characterized by acute severe pneumonia, meningitis, rhinitis and otitis media/interna. The disease is most often reported in shelter cats, but has also rarely been isolated from pet cats [12, 42]. SEZ should be included in differential diagnostic workup of cats with respiratory, neurological or obscure illness with appropriate antibiotic selection for streptococcal pathogens being included in therapeutic regimens. Recent reports suggest that SEZ is associated with chronic rhinitis and otitis media/interna and cats with these conditions might represent the point of introduction into shelters. Thorough physical examination of cats upon introduction to shelters and prompt treatment of rhinitis and/or otitis media/interna with antibiotics effective against streptococcal bacteria may help to prevent outbreaks of acute fatal disease.

References

[1] Greene, C.E. and Prescott, J.F. (2006) Streptococcal infections. In: C.E. Greene, Infectious Diseases of the Dog and Cat (3rd edition, pp. 302-309). St. Louis, Missouri: Saunders.

[2] Wu, C.C., Kiupel, M., Raymond, J.T., de Gortari, M.J. and Lin, T.L. (1999). Group G streptococcal infection in a cat colony. *J. Vet. Diag. Invest.*, 11, 174-177.

[3] Lyskova, P., Vydrzalova, M., Kralovcova, D. and Mazurova, J. (2007). Prevalence and characteristics of *Streptococcus canis* strains isolated from dogs and cats. *Acta Vet. Brno*, 76, 619-625.

[4] Devriese, L.A., Cruz Colque J.I., De Herdt, P. and Haesebrouck, F. (1992). Identification and composition of the tonsillar and anal enterococcal and streptococcal flora of dogs and cats. *J. App. Bacteriol.*, 73, 421-425.

[5] Demko, J. L. and Cohn, L. A. (2007). Chronic nasal discharge in cats: 75 cases (1993-2004). *J. Am. Vet. Med. Assoc.,* 230, 1032-1037.

[6] Johnson, L. R., Foley, J. E., De Cock, H.E.V., Clarke, H. E. and Maggs, D. J. (2005) Assessment of infectious organisms associated with chronic rhinosinusitis in cats. *J. Am. Vet. Med. Assoc.,* 227, 579-585.

[7] Roy, J., Messier, S., Labrecque, O. and Cox, W.R. (2007). Clinical and *in vitro* efficacy of amoxicillin against bacteria associated with feline skin wounds and abscesses. *Can. Vet. J.,* 48, 607-611.

[8] Iglauer, F., Kunstyr, I., Morstedt, R., Farouq, H., Wullenweber, M. and Damsch, S. (1991). *Streptococcus canis* arthritis in a cat breeding colony. *J. Exp. Anim. Sci.*, 34, 59-65.

[9] Tillman, P.C., Dodson, N.D. and Indiveri, M. (1982). Group G streptococcal epizootic in a closed cat colony. *J. Clin. Microbiol.*, 16, 1057-1060.

[10] Sura, R., Hinckley, L.S., Risatti, G.R. and Smyth, J.A. (2008). Fatal necrotising fasciitis and myositis in a cat associated with *Streptococcus canis*. *Vet. Rec.* 162, 450-453.

[11] Pesavento, P.A., Bannasch, M.J., Bachmann, R., Byrne, B.A. and Hurley, K.F. (2007). Fatal *Streptococcus canis* infections in intensively housed shelter cats. *Vet. Pathol.*, 44, 218-221.

[12] Blum, S., Elad, D., Zukin, N., Lysnyansky, I., Weisblith, L., Perl, S., Netanel, O. and David, D. (2010). Outbreak of *Streptococcus equi* subsp. *zooepidemicus* infections in cats. *Vet. Microbiol.* 144, 236-239.

[13] Britton, A.P. and Davies, J.L. (2010). Rhinitis and meningitis in two shelter cats caused by *Streptococcus equi* subspecies *zooepidemicus*. *J. Comp. Pathol.* 143, 70-74.

[14] Chalker, V.J., Brooks, H.W. and Brownlie, J. (2003). The association of *Streptococcus equi* subsp. *zooepidemicus* with canine infectious respiratory disease. *Vet. Microbiol.* , 95, 149-156.

[15] Pesavento, P.A., Hurley, K.F., Bannasch, M.J., Artiushin, S. and Timoney, J.F. (2008). A clonal outbreak of acute fatal hemorrhagic pneumonia in intensively housed (shelter) dogs caused by *Streptococcus equi* subsp. *zooepidemicus*. *Vet. Pathol.* , 45, 51-53.

[16] Byun, J.-W., Yoon, S.-S., Woo, G.-H., Jung, B.Y. and Joo, Y.-S. (2009). An outbreak of fatal hemorrhagic pneumonia caused by *Streptococcus equi* subsp. *zooepidemicus* in shelter dogs. *J. Vet. Sci.*, 10, 269-271.

[17] Garnett, N.L., Eydelloth, R.S., Swindle, M.M., Vonderfecht, S.L., Strandberg, J.D. and Luzarraga, M.B. (1982). Hemorrhagic streptococcal pneumonia in newly procured research dogs. *J. Am. Vet. Med. Ass.* , 181, 1371-1374.

[18] Kim, M.K., Jee, H., Shin, S.W., Lee, B.C., Pakhrin, B., Yoo, H.S., Yoon, J.H. and Kim, D.Y. (2007). Outbreak and control of haemorrhagic pneumonia due to *Streptococcus equi* subspecies *zooepidemicus* in dogs. *Vet. Rec.,* 161, 528-530.

[19] Sundberg, J.P., Hill, D., Wyand, D.S., Ryan, M.J. and Baldwin, C.H. (1981). *Streptococcus zooepidemicus* as the cause of septicemia in racing Greyhounds. *Vet. Med.: Sm. Anim. Clin.*, 76, 839-842.

[20] Timoney, J.F. (2004). Streptococcus. In: C.L. Gyles, J.F. Prescott, J.G. Songer and C.O. Thoen (Eds.), Pathogenesis of bacterial infections in animals (3rd edition, pp. 23-42). Ames, Iowa: Blackwell.

[21] Webb, K., Jolley, K.A., Mitchell, Z., Robinson, C., Newton, J.R., Maiden, M.C.J. and Waller, A. (2008). Development of an unambiguous and discriminatory multilocus sequence typing scheme for the *Strepococcus zooepidemicus* group. *Microbiol.* 154, 3016-3024.

[22] Radostits, O.M., Gay, C.C., Blood, D.C. and Hinchcliff, K.W. (2000). Veterinary medicine: a textbook of the diseases of cattle, horses, sheep, pigs and goats (9th edition). Philadelphia, Pennsylvania: Saunders.

[23] Pusterla, N., Luff, J.A., Myers, C.J., Vernau, W. and Affolter, V.K. (2007). Disseminated intravascular coagulation in a horse with *Streptococcus equi* subspecies *zooepidemicus* meningoencephalitis and interstitial pneumonia. *J. Vet. Int. Med.*, 21, 344-347.

[24] Bailie, W.E., Stowe, E.C. and Schmitt, A.M. (1978). Aerobic bacterial flora of oral and nasal fluids of canines with reference to bacteria associated with bites. *J. Clin. Microbiol.*, 7, 223-231.

[25] Lamm, C.G., Ferguson, A.C., Lehenbauer, T.W. and Love, B.C. (2010). Streptococcal infection in dogs: a retrospective study of 393 cases. *Diag. Pathol.,* 47, 387-395.

[26] Gibson, D. and Richardson, G. (2008). Hemorrhagic streptococcal pneumonia in a dog. *Vet. Rec.*162, 423-424.

[27] Priestnall, S. and Erles, K. (2011). *Streptococcus zooepidemicus:* an emerging canine pathogen. *The Vet. J.*, 188, 142-148.

[28] Oikawa , M., Kamada, M., Yoshikawa, Y. and Yoshikawa, T. (1994). Pathology of equine pneumonia associated with transport and isolation of *Streptococcus equi* subsp. *zooepidemicus*. *J. Comp. Pathol.*, 111, 205-212.

[29] Oikawa, M., Takagi, S., Anzai, R., Yoshikawa H. and Yoshikawa T. (1995). Pathology of equine respiratory disease occurring in association with transport. *J. Comp. Pathol.*, 113, 29-43.

[30] Larson, L.J., Henningson, J., Sharp, P., Thiel, B., Desphpande, M.S., Davis, T., Jayappa, H., Wasmoen, T., Lakshmanan, N. and Schultz, R.D. (2011). Efficacy of the canine influenza virus H3N8 vaccine to decrease severity of clinical disease after cochallenge with canine influenza virus and *Streptococcus equi* subsp. *zooepidemicus*. *Clin. Vacc. Immunol.*, 18, 559-564.

[31] Walker, R.L. and Runyan, C.A. (2003). Identification of variations in SzP proteins of *Streptococcus equi* subspecies *zooepidemicus* and the relationship between protein variants and clinical signs of infection in horses. *Am. J. Vet. Res.*, 64, 976-981.

[32] Anzai, T., Walker, J.A., Blair, M.B., Chambers, T.M. and Timoney, J.F. (2000). Comparison of the phenotypes of *Streptococcus zooepidemicus* isolated from tonsils of healthy horses and specimens obtained from foals and donkeys with pneumonia. *Am. J. Vet. Res.*, 61, 162-166.

[33] Paillot, R., Darby, A.C., Robinson, C., Wright, N.L., Steward, K.F., Anderson, E., Webb, K., Holden, M.T., Efstratiou, A., Broughton, K., Jolley, K.A., Priestnall, S.L., Marotti Campi, M.C., Hughes, M.A., Radford, A., Erles, K. and Waller, A.S. (2010). Identification of three novel superantigen-encoding genes in *Streptococcus equi* subsp. *zooepidemicus*, *szeF*, *szeN*, and *szeP*. *Inf. Immun.,* 78, 4817-4827.

[34] Macdonald, E.S., Norris, C.R., Berghaus, R.B. and Griffey, S.M. (2003) Clinicopathologic and radiographic features and etiologic agents in cats with histologically confirmed infectious pneumonia: 39 cases (1991-2000) *J. Am. Vet. Med. Ass.,* 223, 1142-1150.

[35] Foster, S.F., Martin, P., Allan, G.S., Barrs, V.R. and Malik, R. (2004). Lower respiratory tract infections in cats: 21 cases (1995-2000). *J. Fel. Med. Surg.*, 6, 167-180.

[36] Foster, S.F and Martin, P. (2011). Lower respiratory tract infections in cats, reaching beyond empirical therapy. *J. Fel. Med. Surg.,* 13, 313-332.

[37] Acke, E., Abbott, Y., Pinilla, M., Markey, B.K. and Leonard, F.C. (2010). Isolation of *Streptococcus zooepidemicus* from three dogs in close contact with horses. *Vet. Rec.,* 167, 102-103.

[38] Piva, S., Zanoni, R.G., Specchi, S., Brunetti, B., Florio, D. and Pietra, M. (2010). Chronic rhinitis due to *Streptococcus equi* subspecies *zooepidemicus* in a dog. *Vet. Rec.,* 167, 177-178.

[39] Abbott, Y., Acke, E., Khan, S., Muldoon, E.G., Markey, B.K., Pinilla, M., Leonard, F.C., Steward, K. and Waller, A. (2010). Zoonotic transmission of *Streptococcus equi* subsp. *zooepidemicus* from a dog to a handler. *J. Med. Microbiol.*, 59, 120-123.

[40] Hirst, R.A., Kadioglu, A., O'Callaghan, C. and Andrew, P.W. (2004). The role of pneumolysin in pneumococcal pneumonia and meningitis. *Clin. Exp. Immunol.*, 138, 195-201.

[41] Van Ginkel, F.W., McGhee, J.R., Watt, J.M., Campos-Torres, A., Parish, L.A. and Briles, D.E. (2003). Pneumococcal carriage results in ganglioside-mediated olfactory tissue infection. *Proc. Nat. Acad. Sci.,* 100, 14363-14367.

[42] Martin-Vaquero, P., da Costa, R.C. and Daniels, J.B. (2011). Presumptive meningoencephalitis secondary to extension of otitis media/interna caused by *Streptococcus equi* subspecies *zooepidermicus* in a cat. *J. Fel. Med. Surg.*, 13, 606-609.

[43] Rajasekhar, A. and Clancy, D.J. (2010). Meningitis due to group C Streptococcus: A case report and review of the literature. *Scand. J. Inf. Dis.,* 42, 571-578.

In: Cats
Editors: T. Zakrevsky and V. Lachinov

ISBN 978-1-62081-043-9

Chapter V

FELINE ENDOMETRIAL ADENOCARCINOMAS

Ana Laura Saraiva, Rita Payan-Carreira and Maria dos Anjos Pires
CECAV – University of Trás-os-Montes and Alto Douro, Portugal

ABSTRACT

Endometrial adenocarcinomas are considered to be rare in domestic animals, in particular comparing to non-epithelial tumours of the uterus such as leiomyoma, in part because they are likely underdiagnosed. In cats, though uterus was the most common site for genital tract tumours, endometrial adenocarcinomas were found to be rare in all available references.

In the study presented herein, forty feline primary endometrial adenocarcinomas, identified by a minimum of three pathologists on conventional haematoxylin and eosin-stained sections, obtained from the archives of four different laboratories, for a period of 13 years were used in a retrospective study on the subject.

In previous studies on feline endometrial adenocarcinomas (FEA) it has been found an increase incidence in purebred animals, which has been associated with a longer reproductive activity in those animals in comparison with domestic cats. However, this should be read with caution, as population ratios between purebred and domestic mongrel animals may change with the geographic location. Also the cultural adoption of gonadectomy as a contraceptive measure may interfere with the regional prevalence of the disease. In fact, in Portugal the number of intact domestic cats is higher than the purebreds, due to the tendency for late spaying age (most of them due to uterine or mammary diseases), which could influence the incidence of FEA in our country.

Some of the specimens used in this study were obtained at ovariohysterectomy following a clinical diagnosis of pyometra. Unless metastasic disease exists, to which illness is often associated, most frequent signs for FEA are vague and unspecific, making difficult the early diagnosis of endometrial carcinoma.

The histological evaluation shows the existence of different tumour phenotypes, where clear cell are often found. The presence of nuclear atypia and the Ki67 expression seem not to be related to the malignancy of the tumour. Furthermore, one report refers the presence of oestrogen receptors, which could explain the low levels of metastization or invasiveness proposed for FEA, or the clinical outcome for the tumour. For most situations of primary FEA, in particular in early stages of the disease, myometrial invasion is limited and often absent. Also vascular invasion is seldom observed even in tumours with moderate nuclear atypia. However, limited information exists on specimens from early FEA stages, which impair the presentation of a reliable prognosis for that animal. The major challenge seems to be the identification of valuable prognostic markers to achieve a definitive prognosis for animal life.

Feline endometrial adenocarcinomas are possibly more common than we might presume and the clinical impact of such tumours may be increasing with the increase in the ability to establish an early diagnosis allied to an extended duration of life in cats. A recent report defends that FEA are more common than smooth muscle tumours of the uterus. Furthermore, as it frequently co-exists with pyometra or mucometra, it is also possible that a large number of cases fail to reach the pathological evaluation and hence remain undiagnosed.

In this chapter we propose to discuss the clinical and morphological data of all the forty feline primary endometrial adenocarcinomas, supported by an extended review of the literature

1. Introduction

Although the uterus is considered the most common site of feline reproductive tract tumours in one study, uterine tumours comprised only 0,29% of all feline neoplasms diagnosed during the study period (Miller et al., 2003). Adenocarcinomas of the endometrium are considered to be rare in all domestic species, with exception of cattle and rabbits (Cotchin, 1964; Kennedy et al., 1998; McEntee and Nielsen, 1976), with the common practice of ovariohysterectomy in cats being referred as protective from uterine neoplasia (Miller et al., 2003; Taylor, 2010). Spaying of young queens for contraception is neither a worldwide procedure nor a common practice in purebred cats, and hence is may not always be the reason for the small percentage of uterine adenocarcinomas. Furthermore, there are also reports of adenocarcinoma in the uterine stump of neutered cats (Anderson & Pratschke, 2011; Miller et al., 2003). On the other hand, the rarity of these tumours in cats might be related to inadequate post-mortem examination of the genital tract, but also to lack of interest in anatomopathological evaluation of ovariohysterectomy surgical specimens. In fact, in the study by Miller et al. (2003), from all the uterine tumours, the endometrial adenocarcinoma was the most commonly found in cats.

According to clinical reports and our own experience, uterine carcinoma does not present specific signs. It rather shares some clinical signs with other uterine diseases and abortion, such as fluid accumulation with distension of the organ and increased thickness of the uterine walls, combined with emaciation or vaginal haemorrhagic discharge (Preiser, 1964; Belter et al., 1968; Anderson & Pratschke, 2011). These features further strength the possibility that the situations may be underdiagnosed, as the excised genital segments are seldom sent to histopathological analysis. Endometrial adenocarcinoma occurs most frequently in middle age to old animals, and its putative association with progestagens-based contraception is controversial. Nevertheless, very young animals can develop the disease.

In other species, endometrial adenocarcinoma evolves as a malignant epithelial neoplasm. Also in felids, some reports exist on the presence of metastasis at the moment of diagnosis (Preiser, 1964; Belter et al., 1968; Anderson & Pratschke, 2011); however, its occurrence seems sporadic, and predictive markers have not yet been established.

The major challenge seems to be the identification of diagnostic markers of feline endometrial adenocarcinomas, since there are few reports about these features, all of them in small case series. Because of the same reason, the achievement of prognosis of these lesions is difficult and requires further studies.

Different morphologies of feline endometrial adenocarcinomas can be defined, which included papillary serous carcinoma (with or without clear cells), clear cell carcinoma and in situ carcinoma according to cytological criteria of malignancy, cell morphology and invasion of adjacent tissues (Saraiva et al., 2011).

It is the intent of this paper to summarize what is known about the feline endometrial adenocarcinoma, supported by the findings of 40 histopathological situations. We will discuss the incidence of the disease, the main clinical signs as well as its diagnosis and the need for reliable prognostic markers. Further we will characterize the main morphological features of the three main types of feline endometrial adenocarcinoma (FEA) found in our casuistic.

2. Incidence/Epidemiology of FEA

Endometrial carcinoma/adenocarcinoma is more common in the cat than the dog (Morris &Dobson, 2001). Uterine adenocarcinoma is believed to be locally invasive, and the occurrence of metastasis are commonly reported. Severe illness is mainly associated with metastatic disease and is often at the origin of the presentation at the clinic.

Although no breed predisposition has been reported (Klein, 2007), in previous studies on feline endometrial adenocarcinomas (FEA) it has been found an increase incidence in purebred animals, which has been associated with a longer reproductive activity in those animals in comparison with domestic cats (Johnston et al., 2001). However, this should be read with caution, as population ratios between purebred and domestic mongrel animals may change with the geographic location. Uterine samples with FEA were more frequently found, within the cases used herein, in domestic animals than in purebred. This might be explained by the fact that in Portugal the number of intact domestic cats is higher than the purebreds, due to the tendency for late spaying age (most of them due to uterine or mammary diseases), which could influence the incidence of FEA in our country.

From our database, in the last thirteen years, 197 female genital tracts have been sent for analysis (almost 12% of all female feline analysis), from which 41.16% (81:197) didn't show signs of uterine disease, 6.6% (13:197) were pregnant or with signs of abortion, 33.5% (66:197) showed inflammatory diseases of the uterus and 20.30% (37:197) were diagnosed with FEA. The relative frequency of FEA is quite high in our database, which could be associated to the fact that in the last three years, we performed the systematic evaluation of excised genitalia for a PhD project. The analysis of the frequencies for the other clinical situations also highlights the fact that most frequently the clinicians send to commercial laboratories for analysis the organs with macroscopic signs of disease. Consequently, it is

possible that small lesions may be undetected during their first trial, hence not being submitted for diagnosis, or incorrectly assumed as having pyometra.
Furthermore, feline adenocarcinoma was most frequently found in domestic cats than in purebred (with the registry of only two Siamese and one Persian queen).

FEA most frequently develops in older animals, with more than 9 years (Preiser, 1964; Belter et al., 1968; McEntee, 1990; Morris & Dobson, 2001-Small Animal oncology; Klein, 2007 – Small Animal clinical oncology), although sporadic reports on younger animals around 5 years old are published. Although clinical information was only possible for half the animals used in our studies, in our 40 cases the age of the females diagnosed with FEA ranged from 2 to 15 years old. The age mean was 7 years and median for age was 6 years old.

3. Clinical Signs

Albeit several studies on feline endometrial carcinoma have been published, few reports detail the clinical aspects of the pathology. Furthermore, most descriptions refers to late stages of the disease (Preiser, 1964; Belter et al., 1968; Anderson & Pratschke, 2011), as often FEA evolves as a silent disease, not detected clinically unless large lesions or metastases are present. Illness is usually related to metastatic disease, and the major presenting complaint.

The clinical signs of FEA vary with the size of the neoplasic lesions, but also with the age of the process and the existence and pattern of metastasis. It is possible that the morphological type of the tumour may also interfere with the most prevalent clinical sign. The size of the tumour may vary from less than one centimetre to greater than 10 cm (Johnston et al., 2001). Small-sized tumours are most frequently found incidentally at ovariohysterectomy or post-mortem, whilst the large one usually gives more or less notorious clinical signs.

If metastases are not present, vague and unspecific signs of uterine disease may be found, since the presence of a mass within the uterus may induce in chronic inflammation causing a mucous, purulent or haemorrhagic vaginal discharge, or pyometra (Taylor, 2010). Vulvar discharge may be intermittent (Meier, 1956, cited by McEntee, 1990). As consequence of this inflammatory reaction signs such as vomiting, irregular appetite and polyuria/polydipsia are present, and together with the vaginal discharge are usually the main complaints at consultation. Rarely, abnormal oestrous cycles have been described in association with FEA. Enlargement of the uterus, abdominal distress and distension of the abdomen are also frequently found during transabdominal palpation or when larger tumours exist (Johnston et al., 2001; Taylor, 2010). Also, when the lesions grow large enough to compress adjacent viscera, other signs may develop (Klein, 2007), like constipation or dysuria. They are frequently found in published reports on metastatic FEA situations, but are not always associated with the metastasis itself. Whenever metastasis exists, ascites, anorexia and weight loss are commonly found clinical signs, which may co-exist with anaemia, fever and cachexia (Meier, 1956, cited by McEntee, 1990; Johnston et al., 2001). FEA may metastasize regionally or at the distance, to lungs, brain, or eyes (McEntee, 1990; Klein, 2007; Taylor, 2010). Depending on the location of the metastasis, secondary symptoms may be found: cough and respiratory distress, if metastases are thoracic, blindness or motor incoordination if in the central nervous system, ascites if in the abdominal organs. According to Taylor (2010) most cats develop metastatic disease within 6 months of diagnosis, if uterine excision is not performed or lymphatic invasion already exist at diagnosis. However, this period should be considered as indicative, as for most published reports, the tumour was diagnosed in later stages of the disease.

4. Morphological Features of Primary Epithelial Tumours of the Uterus

Several studies had been published in small series of feline endometrial adenocarcinomas, mainly related with their immunoprofile (Espinosa de los Monteros et al., 1999; Gil da Costa et al., 2009; Martin de las Mulas et al., 1995; Miller et al., 2003), but there is lack of literature explaining the different morphologic aspects of these tumours, even when case reports are described (Belter et al., 1968; Sapierzynski et al., 2009; Anderson & Pratschke, 2011).

Based in our experience, there are three different morphologies of primary endometrial adenocarcinomas in cats, which can be compared with women endometrial carcinoma (Horn et al., 2007): papillary serous carcinoma, clear cell carcinoma and a lesion restricted to the superficial endometrium that we might classified as "in situ" carcinoma.

Macroscopically, these three different morphologies appear to be indistinguishable, presenting as a diffuse thickening of the endometrium along both uterine horns and corpus or as multiple white nodules, projecting to the lumen (Figure 1). The uterine wall can be also thinner if an abundant purulent content is present within the organ. In invasive tumours, it is possible to detect tumoural infiltration through the myometrium, sometimes with umbilication in the serosa that could rupture, promoting in this cases peritonitis.

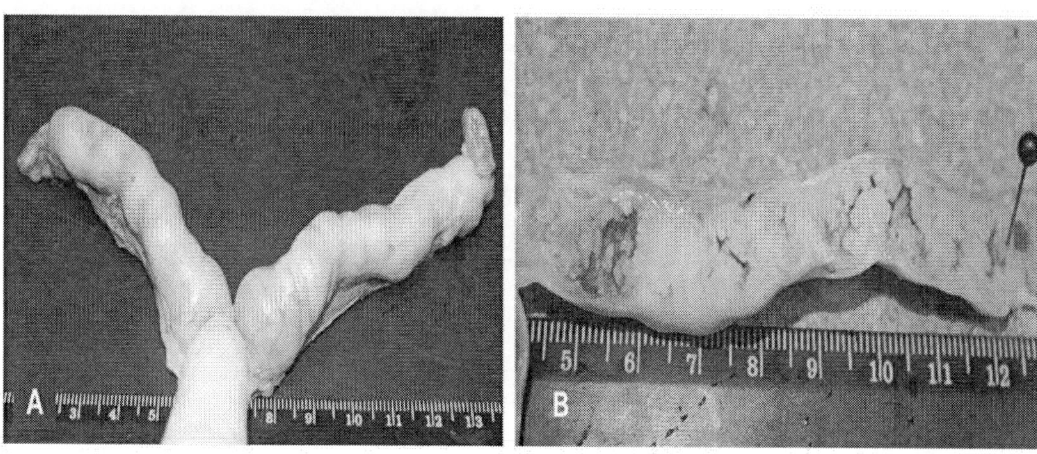

Figure 1. Feline endometrial adenocarcinoma. A - The uterus is enlarged in all the extension. B - The endometrium is diffusely thicker through the uterine horn, appearing as a proliferative white tissue.

Papillary Serous Carcinoma

The majority of feline endometrial carcinomas, as described in the references cited above, have a papillary growth pattern projected into the lumen (Figure 2A).

The papillae are supported by a thin fibrovascular stroma (that may be edematous or hyalinized) and lined by more than one layer of neoplastic cells. The cells are pleomorphic columnar shaped, with moderate amount of eosinophilic cytoplasm, sometimes with focal clearing. The nuclei are round to oval, vesicular or hyperchromatic, and loose the normal polarity; large nucleoli are evident and sometimes are found intranuclear clear inclusions. Mitotic figures are frequent and frequently bizarre. There is a moderate degree of anisokaryosis and anisocytosis. Numerous multinucleated cells can be found within the tumour. In some tumours, is possible to see squamous metaplasia. Randomly distributed areas of necrosis are found within the neoplasm and occasionally psammoma bodies are present. The invasion of the myometrium, vessels and lymphatics is not a constant feature. Some tumours have a clear inflammatory component, with macrophages, plasma cells, lymphocytes, neutrophils and occasionally eosinophils. Although these tumours show

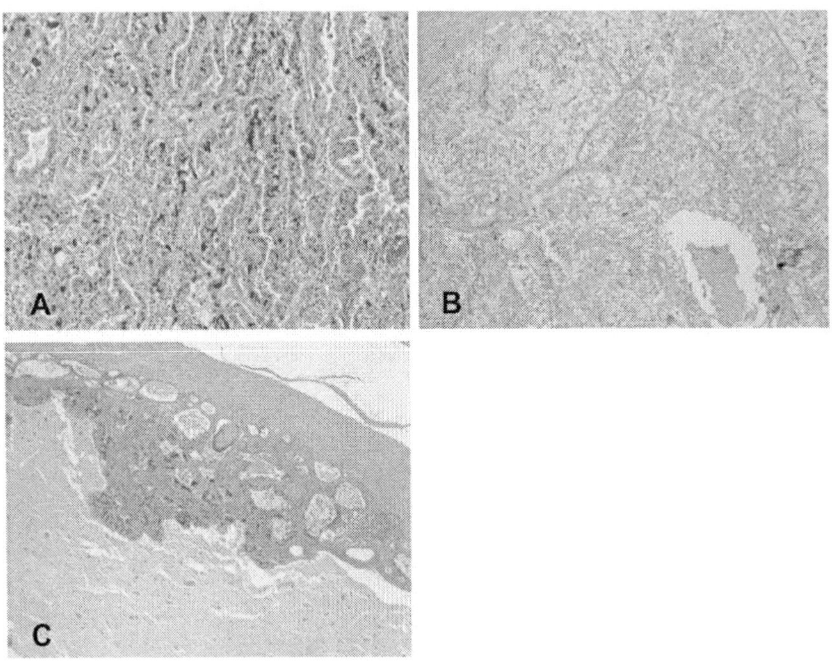

Figure 2. Morphological aspects of feline endometrial adenocarcinomas (HE). A. Papillary serous carcinoma (Obj. 10x). The papillae are surrounded by more than one layer of neoplastic cells and supported by scant fibrovascular stroma. It is possible to observe several multinucleated cells. B. Clear cell carcinoma (Obj. 10x). Nests of foamy cells, sometimes with central necrosis, surrounded by fibrovascular stroma. C. "In situ" carcinoma (Obj. 4x). Non invasive, superficial malignant proliferation of the endometrium.

a predominant papillary serous morphology, areas of solid growth and a glomerular pattern may be present as well as various numbers of clear cells, which are described in the next section.

Clear Cell Carcinoma

Clear cell carcinomas of the feline endometrium are almost entirely composed by large, round to polygonal cells, with foamy cytoplasm and eccentric crenate or ovoid nucleus, with a prominent eosinoplilic nucleolus. The cells are arranged in papillae, sheets or solid nests surrounded by a fibrovascular stroma (Figure 2B). As in papillary serous carcinoma, there is a moderate degree of anisokaryosis and anisocytosis but multinucleated cells are absent. It can also be found necrosis and inflammatory cells are rare. The invasion of the myometrium is not a constant feature. Vessel or lymphatic tumoral emboli were not found in the available cases.

"In Situ" Carcinoma

Morphologic aspects of "in situ" carcinoma are very similar with those described for the papillary serous carcinoma. The differences are related to the location of this tumour that is very superficial in the apical endometrium and to the lack of invasion of the stroma (Figure 2C). As the name implies, these tumours are non invasive.

5. Diagnostic Markers

Major available reports do not distinguish between the three above-mentioned FEA phenotypes when studying the expression of molecular markers in uterine epithelial tumours. However, it is possible that their grow, invading behaviour and metastatic potential may differ between the tumour′ types. Only by searching for the differences we may be able to develop markers that can be use for treatment and prognosis.

Adenocarcinomas of the endometrium, like other epithelial tumours, tend to express broad-spectrum cytokeratins (CK), clones AE1/AE3 and MNF 116, but reactivity with AE1/AE3 is generally more intense and more frequent than with MNF 116 (Gil da Costa et al., 2009; Miller et al., 2003; Sapierzynski et al., 2009). Espinosa de los Monteros et al., 1999 described the coordinate expression of cytokeratins 7 and 20 in 67% of the feline endometrial carcinomas under study, in accordance with the results observed in the normal feline endometrium. Therefore, they proposed that coordinate expression of these two cytokeratins provides important information that can help in the diagnosis of these tumours (Espinosa de los Monteros et al., 1999). The tumours do not express CK14 (Gil da Costa et al., 2009), as well as intermediate filament cytokeratins RDK-102 and NCL-5D3 (Martín de las Mulas et al., 1995).

The great majority of feline endometrial carcinomas are negative for vimentin, but there are reports of tumours with expression of this molecule (Gil da Costa et al., 2009; Martín de las Mulas et al., 1995; Miller et al., 2003).

According to the literature, feline endometrial adenocarcinomas are generally negative for the estrogen receptor-α (ERα), but in some tumours, up to 50% of the neoplastic cells have nuclear staining for ERα (Gil da Costa et al., 2009; Miller et al., 2003; Sapierzynski et al., 2009). Miller et al., 2003 found a tendency for metastases by the tumours that are negative for ERα and they defend that the loss of expression of estrogen receptors in such tumours suggests estrogen independence and may indicate a worse prognosis.

The score of expression of progesterone receptor is very variable in endometrial adenocarcinomas, as well as the intensity of expression of this marker, but there is a tendency to a markedly reduced expression in these neoplasms (Gil da Costa et al., 2009; Sapierzynski et al., 2009).

There is a moderate to strong expression of cyclooxygenade-2 (COX-2) by up to 50% of the neoplastic cells and a loss of polarity compared with the normal endometrium (Gil da Costa et al., 2009).

Disperse information on other molecular markers, not explored from the biological point of view on respect to the tumour behaviour, are available in the literature. Endometrial adenocarcinomas of cats express E-cadherin and β-catenin, as normal endometrium (Gil da Costa et al., 2009; Sapierzynski et al., 2009). In one case report, about 40% of tumoral cells of adenocarcinoma epithelium stained with Ki67 (Sapierzynski et al., 2009). The tumours are negative for smooth muscle actin, desmin, glial fibrillary acid protein and caveolin-1 (Gil da Costa et al., 2009; Martín de las Mulas et al., 1995; Miller et al., 2003; Sapierzynski et al., 2009).

6. Clinical Diagnosis

History, presenting complaint, and physical examination findings usually provide the basis for diagnostic. Vaginal bleeding or vaginal discharge may direct the differential diagnosis into abortion, uterine tumour or cystic endometrial hyperplasia/pyometra.

On early stages of the disease, FEA progress silently or giving discrete clinical signs. Most frequently, when signs of uterine disease are present, the animal is submitted to ovariohysterectomy (OVH) under the suspicion of cystic endometrial disease or pyometra. Thus, the diagnosis is normally made at OVH or at necropsy. No consistent laboratory abnormalities have been noted, unless pyometra exist, and abdominal radiographs only confirm the presence of an abdominal or uterine mass. Hematological abnormalities might include regenerative anemia or neutrophilia (Morris &Dobson, 2001; Klein, 2007).

Whenever clinical signs for uterine disease are present, the diagnosis is based on the clinical signs, abdominal palpation and ultrasound or radiological exploration of the abdominal and thoracic cavities. A soft tissue mass in the mid or caudal abdomen may be detected on the palpation of abdomen, and its echogenicity may be study by plain abdominal radiography or ultrasonography (US). US findings may include hypoechoic masses in one or both uterine horns or lesions miming CEH lesions. As frequently accompanying mucometra or pyometra are found, enlarged coils of uterus may be distinguishable, along with the presence of small amount of fluid in the uterus. Ultrasonography may further be useful to define the origin of the mass or to evaluate regional metastasis (Morris &Dobson, 2001; Klein, 2007). Sporadically, surgical exploration may be needed to achieve the diagnosis (Klein, 2007; Taylor, 2010). Radiographic findings can include soft tissue density masses that may displace the surrounding viscera. A definitive diagnosis is only obtained through histological examination of surgically excised specimens (Klein, 2007).

In FEA more advanced stages, clinical signs may occur as a result of metastatic disease even before any clinical indication of the primary tumour is detected. In such cases, the overall clinical signs may delude the origin of the carcinoma. If ascitis is detected, peritoneal fluid can be collected via abdominocentesis and sent for analysis.

Survival and outcome data are sparse metastatic disease due to FEA (Klein, 2007; Taylor, 2010). The poor animal condition at diagnosis frequently limits the chances for successful treatment and often euthanasia is the human outcome for it. The differential diagnoses should contemplate other metastatic diseases associated to other carcinoma, as those at the mammary glands, liver, etc. In the absence of neurological signs, the presence of metastases on the lungs, eye, abdominal lymph nodes, liver and diaphragm should be ascertain, as it can influence both the treatment and the prognosis.

7. TREATMENT

In case of co-existing pyometra, supportive care should be provided. The use of antibiotics has to be pondered individually.

If metastatic disease is still not apparent, a complete OVH is recommended, and the uterus should be sent (the entire surgical piece) for histopathological analysis. Whenever the uterus shows abnormal macroscopic features, even if gross evidence of a tumour is not present, the histopathological analysis ought to be requested. During the surgical procedure, the abdominal cavity should be explored and any suspected mass biopsied for histopathology. Attempts should be made to remove all tumours and metastatic foci.

Chemotherapy and radiation therapy for treating uterine tumours are largely untested in veterinary cases. However, adjunctive chemotherapy may be considered if malignancy is detected upon histopathology (Morris &Dobson, 2001). Its use may be of survival benefit for patients, as it may prevent or delay the development of metastasis. However, the owner must be advertised that the benefits are yet unclear. For cats, the use of doxorubicin or carboplatin has been suggested. As limited information on the chemotherapy outcome for uterine tumours, health parameters should be closely monitored for undesirable side effects. Follow-up appointments should be schedule every three months to check for cancer spread and adjust the therapy as needed.

8. Prognosis

In general, because clinical signs rarely are evident until late in the course of disease, the prognosis should always be considered guarded, in particular because FEA have a documented metastatic potential. Metastatic disease warrants a grave prognosis.

Nevertheless, the quality of the studies made so far about this theme doesn't allow the accurate determination of prognosis of endometrial adenocarcinomas in cats, and its metastatic potential depends on the number of clinical reports, but not on trustful clinical surveys. Only one study reveals that there might be a tendency to metastases by the tumours that are negative for ERα, which can indicate a worse prognosis (Miller et al., 2003). In the case reports available in the literature, the tumours presented in an advanced stage, resulting in a reserved prognosis and euthanasia (Belter et al., 1968; Anderson & Pratschke, 2011; Preiser, 1964; Sapierzynski et al., 2009). According to our experience, if the tumour is detected precociously and the ovariohysterectomy is performed, the prognosis is favourable. Furthermore, "in situ" carcinomas, non invasive papillary serous carcinomas and non invasive clear cell carcinomas tend to have a better prognosis in the immediate future, but a longer period of time should be evaluated for definitive conclusions.

Conclusion

Feline endometrial adenocarcinomas are possibly more common than the literature describes and distinct morphologies are here proposed: papillary serous carcinoma, clear cell carcinoma and "in situ" carcinoma. Few studies are available in small case series of these tumours, which difficult the determination of several aspects of these neoplasms. Nevertheless, all of them describe the loss of expression of estrogen receptor by neoplastic cells, as being a consensual feature of feline endometrial adenocarcinomas, and possibly related to a worse prognosis. So far, from other authors' studies, this seems to be the best diagnostic marker associated with prognosis for feline endometrial adenocarcinomas. Another feature of feline endometrial adenocarcinomas is the coordinate expression of cytokeratins 7 and 20, which helps the diagnosis of these tumours. COX-2 may play an important role in the carcinogenesis of the endometrium, but more studies in larger case series are expected to extend the existent knowledge. New studies including the clinical data on the patients diagnosed with FEA in early stages of the disease are desired in order to establish incidence, common clinical profiles and survival rates for the disease.

Acknowledgments

The authors thank to Ms. Ligia Bento and Alexandra Rema for technical support.

This work was supported by the project from CECAV/UTAD with the reference PEst-OE/AGR/UI0772/2011, by the Portuguese Science and Technology Foundation.

REFERENCES

Anderson C., Pratschke K., 2011. *Uterine adenocarcinoma with abdominal metastases in an ovariohysterectomised cat.* Journal of Feline Medicine and Surgery 13, 44-47.

Belter L.F., Crawford E.M., Bates H.R., 1968. *Endometrial Adenocarcinoma in a Cat.* Pathologia Veterinaria 5, 429-431.

Cotchin E., 1964. *Spontaneous Uterine Cancer in Animals.* British Journal of Cancer XVIII(2), 209-227.

Espinosa de Los Monteros A., Fernández A., Millán M.Y., Rodríguez F., Herrárez P., Martín de las Mulas J., 1999. *Coordinate Expression of Cytokeratins 7 and 20 in Feline and Canine Carcinomas.* Veterinary Pathology 36, 179-190.

Gil da Costa, R.M., Santos M., Amorim I., Lopes C., Dias Pereira P., Faustino A.M., 2009. *An Immunohistochemical Study ofFeline Endometrial Adenocarcinoma.* Journal of Comparative Pathology 140, 254-259.

Horn L-C., Meinel A., Handzel R., Einenkel, J., 2007. *Histopathology of endometrial hyperplasia and endometrial carcinoma – An update.* Annals of Diagnostic Pathology 11, 297-311.

Johnston, S.D., Root Kustritz, M.V. and Olson, P.S. 2001. *Disorders of the feline uterus and uterine tubes (oviducts).* In: Canine and Feline Theriogenology, Saunders/Elsevier, Philadelphia, USA, 463-471

Kennedy, P.C., Cullen, J.M., Edwards, J.M., Goldshmidt, M.H., Larsen, S., Munson, L., Nielsen, S., 1998. *Histological Classification of Tumors of the Genital System of Domestic Animals.* Armed Forces Institute of Pathology in cooperation with the American Registry of Pathology and The World Health Organization Collaborating Center for Worldwide Reference on Comparative Oncology, Washington, DC, pp. 32.

Klein, M.K. 2007. *Tumours of the female reproductive system: uterine tumours.* In: Withrow and MacEwen's Small Animal Clinical Oncology (4rd Ed), Withrow, S and Vail, D (Ed.), Saunders/Elsevier Health Sciences, St Louis, Missouri, USA, pp 613-614.

Martín de las Mulas J., Espinosa de los Monteros A., Carrasco L., van Niel M., Fernández A., 1995. *Immunohistochemical Distribution Pattern of Intermediate Filament Proteins in 50 Feline Neoplasms.* Veterinary Pathology 32, 692-701.

McEntee K., Nielsen S.W., 1976. XVI. *Tumours of the female genital tract.* Bulletin of the World Health Organization 53, 217-226.

McEntee, K. 1990. *The uterus: atrophic, metaplasic and proliferative lesions – neoplasia.* In: Reproductive Pathology of Domestic Animals, Academic Press, San Diego, California, USA, pp 179-190.

Miller M.A., Ramos-Vara J.A., Dickerson M.F., Johnson G.C., Pace L.W., Kreeger J.M., Turnquist S.E., Turk J.M., 2003. *Uterine neoplasia in 13 cats.* Journal of Veterinary Diagnostic Investigation 15, 575-522.

Morris, J. and Dobson, J.M. 2001. – *Genital tract: uterus and cervix.* In: Small Animal Oncology, Wiley-Blackwell, Oxford, London, UK, pp 169-171.

Preiser H., 1964. *Endometrial Adenocarcinoma in a Cat.* Pathologia Veterinaria 1, 485-490.

Sapierzynski R.A., Dolka I., Cywinska A., 2009. *Multiple pathologies of the feline uterus: a case report.* Veterinarni Medicina 52(7), 345-350.

Saraiva A.L., Payan-Carreira R., Gärtner F., Pires M.A., 2011. *C5- Morphological Aspects of Feline Endometrial Adenocarcinomas: a preliminary study.* Abstracts Book of the II Iberic Meeting of Veterinary Pathology, 27.

Schlafer, D.H. and Miller, R.B., 2007. *Female genital system.* In: Jubb, Kennedy and Palmer's Pathology of Domestic Animals, Volume 3, Fifth Ed. Saunders/Elsevier, Philadelphia, PA, USA, pp. 549.

Taylor, K.H. 2010 *Female Reproductive Tumours. In:* Cancer management in small animal practice [Part V: Specific tumours], Henry, C.J. and Higginbotham, M.L., Saunders/Elsevier Health Sciences, St Louis, Missouri, USA, pp: 268-274.

In: Cats
Editors: T. Zakrevsky and V. Lachinov

ISBN 978-1-62081-043-9

Chapter VI

NONSPATIAL VISUAL RECOGNITION MEMORY IN CATS

V. M. Okujava and T. A. Natishvili
Research Center for Experimental Neurology, Tbilisi, Georgia

As it is well known in experimental psychology the evaluation of memories and their characteristics in people are produced by two basic methods: by the method of "free recall" and by the method of "recognition". As far as we know, for evaluating the memory in animals thus far effectively is used only the second of those methods. Since this part is dedicated particularly to memory in subhuman animals, speech subsequently will concern in essence precisely only the recognition memory, with small exceptions. At the same time we were deeply impressed by the success obtained by neuropsychologists in the USA in the study of recognition memory in animals (mainly in monkeys and rats), especially by the work of M. Mishkin's team at Bethesda laboratory of Neuropsychology.

The selection of cats as experimental animals for investigating the memory of recognition was caused in essence by the following bases: 1. As far as we know the recognition memory was investigated mainly in the primates and rats; therefore it appeared to be interesting to establish its presence or absence in the cats as representatives of carnivores, in particular if we consider the fact that the cat, until now, remains classical object for the neurophysiological studies.

Delayed matching- and nonmatching-to-sample (DMS and DNMS) tasks have been widely used to study one-trial object recognition in monkeys [Gaffan, 1974; Mishkin, Delacour, 1975; Mishkin,1982] and to elucidate the cerebral mechanisms that mediate this type of cognitive visual memory [Mishkin,Murray, 1994; Squire, 1992]. Similar studies have been carried out in rodents [for review see Steckler, Drinkenburg, Sahgal, Aggleton, 1998] and other mammals, such as dogs [Callahan, Ikeda-Douglas, Head, Cotman, Milgram, 2000]. But to our knowledge there are no such reports on cats, although some researchers have shown that cats have good memory for spatial locations as measured by their performance on the classical spatial delayed response task [Beritashvili, 1971; Fletcher, 1965; Konorski1967; Warren, Warren, Akert, 1972]]. The purpose of the present part is twofold – one is to uncover the possibility of normal cats to solve different versions of nonspatial delayed matching- and nonmatching-to-sample tasks and the other – to uncover the possible brain regions perhaps responsible for the nonspatial visual memory in cats.

We believe that the most satisfactory neurobehavioral studies addressing interactions between the brain and behavior have until recently been performed in primates. On the other hand, the preferred study system for purely neurophysiologic studies has to date been the cat. Both fundamental and comparative studies with a wide evolutionary perspective need these two approaches to be integrated. The major point here is that the whole arsenal of state-of-the-art neurophysiologic methods can be used in cats for identification of the basic neuronal mechanisms of recognition memory. With this purpose we elected to study in cats some of the forms of cognitive memory which have been well studied in monkeys. In particular, this relates to visual recognition memory in conditions of one-trial visual recognition, which has previously been studied in the USA [Mishkin, Delacour, 1965; Mishkin, Prockop, Rosvold, 1962] using original methods, i. e., image-based delayed selection. Results of such study already have been presented by us earlier. As regards the cerebral mechanisms of this type of visual memory, Murray, Bachevalier, Mishkin [1989] and Meunier, Bachevallier, Mishkin, Murray [1977] showed that the perirhynal and entorhinal cortical areas in primates together (i. e. the “rhinal area”) play the key role in the organization of this type of memory. Is the link between this type of memory and the rhinal area a characteristic of primates or is it more widespread among mammals? As it has been demonstrated by other investigators in dogs [Callahan, Ikeda-Douglas, Head, Cotman, Milgram, 2000] and by us in cats [Okujava, Natishvili, Mishkin, Gurashvili, Chpashvili, Bagashvili, Andronikashvili, Kvernadze, 2005] as representatives of carnivores they can solve the visual nonspatial recognition memory tasks. On the other hand studies in rats have demonstrated not only the existence of visual nonspatial memory, but also the involvement of the rhinal cortical areas in performing the corresponding tasks [Mumby, Pinel, 1997].

In the context of neuropsychological studies of memory, our specific intent was to use the “delayed matching-to-sample” and “delayed nonmatching-to-sample” tests to determine the contribution made by these cortical areas in cats to one-trial non-spatial visual recognition memory.

Materials and Methods

Experimental Subjects

Fifteen experimentally naïve adult normal cats of both sexes (eight male, seven female) weighing 3-4.7 kg were used in this study. The animals were housed in individual cages (1.5 x 1.0 x 1.0 m) in which they had free access to water. Food was given once daily, 20 h before testing. Experimental sessions were conducted 5 days per week. The care and use of the animals complied with Georgian regulations, with Guidelines prepared by the Ethics Committee of the Institutional Animal Care and Use Committee of the Research Center for

Experimental Neurology, and with the National Institutes of Health Guide for the Care and Use of Laboratory Animals.

Five different tasks were used, two requiring manipulatory responses, and three – locomotor responses.

Task I: M-DMS (Manipulatory Variant of DMS)

Four cats, two of them males, were trained on this task in modified Wisconsin General Testing Apparatus (WGTA).

Apparatus

The Wisconsin General Testing Apparatus (WGTA) was adapted for use with cats so that they could use their forelimbs to displace objects and retrieve food. The apparatus as described in detail by Okujava, Natishvili, Mishkin, Gurashvili, Chpashvili, Bagashvili, Andronikashvili, Kvernadze [2005] consisted of two main parts (figure 1): A cat start-cage (55cm x 65cm x 60 cm) placed on a table inside a darkened, sound-shielded room; and a test tray containing three identical food wells, each a round glass jar (25 mm deep and 73 mm in diameter), on which different objects (stimuli) may be placed. The stimuli consisted of an array of 600 junk objects, which differed from each other in size, form, texture, and color (the latter providing mainly brightness cues for cats).

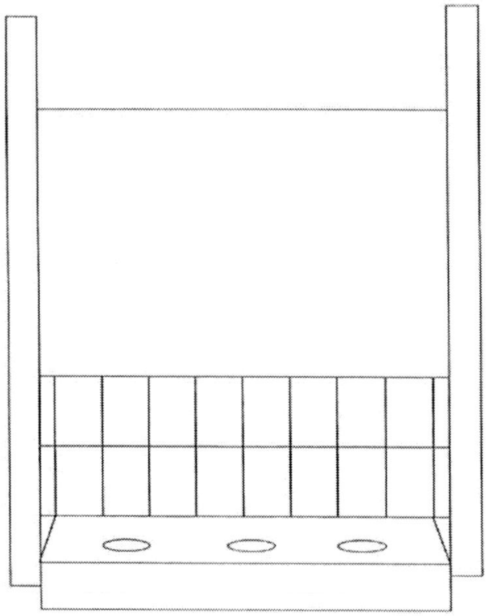

Figure 1. Schematic sketch of the WGTA-type apparatus used for tasks I and II. The side of the cat cage facing the test tray consists of metal bars. The tray, located 15 cm in front of the bars, contains three food wells 15 cm apart (center to center). The opaque screen between the cage and the tray is shown in the raised position.

Procedures

During *preliminary training*, cats were shaped behaviorally to displace cardboard covers placed over the three food wells to obtain rewards (each a small piece of boiled meat, 0.5 cm^3) hidden in the wells. They were then trained in the same way to displace one of three pretraining objects, which were presented singly in random order over one of the three food wells. Finally, the cats were given 20 pseudotrials to familiarize them with the structure of the task: one of the three pretraining objects was presented as the "sample" object over the baited central well; 10 s later the two other objects were presented over the lateral wells, both or neither of which were baited, in random order. The cat was allowed to displace only one of the two "test" objects. The pseudotrials were separated by 30 s intervals. During the 10 s

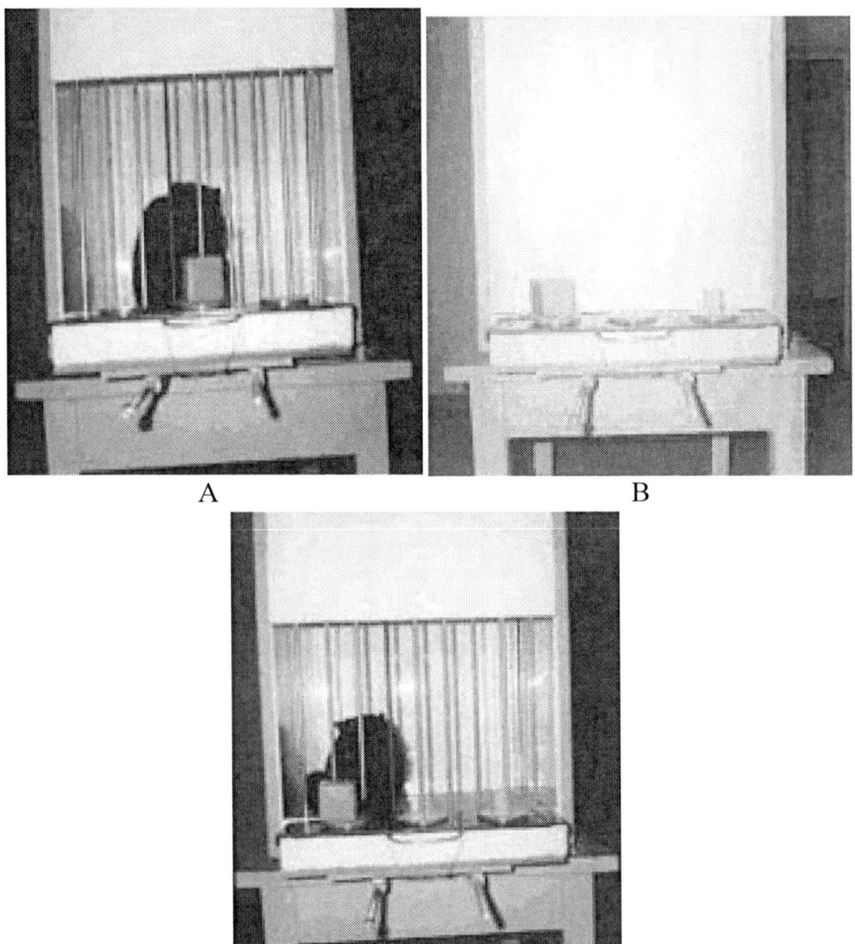

A B

C

Figure 2. Successive stages of the "delayed matching to-sample" (DMS) test procedure is shown below on the photo of the apparatus with cat. A – sample presentation phase; B – delay phase; C – choice phase.

delay intervals and the 30 s intertrial intervals, an opaque screen separated the cat from the test tray. This preliminary training was completed in 7-12 days. *Formal testing* was then begun, using trial-unique objects.

Task I: M-DMS (Manipulatory Variant of DMS) Task

Each trial consisted of two parts, a sample presentation followed by a choice test (figure 2).

After the animal displaced the sample object from the central well and retrieved the reward (no other object was on the test tray), the opaque screen was lowered for delay intervals of 5 and 10 s in pseudorandom order. The screen was then raised revealing the sample object again together with a novel object, each covering one of two lateral wells, and the cat was allowed to choose. A new pair of objects was used on every trial, and the left-right positions of the sample and novel objects on the choice test varied pseudorandomly. In the choice tests of this task, the sample object was always baited, requiring the animal to learn the rule of delayed matching-to-sample. Twenty such trials, separated by 30 s intertrial intervals, were presented daily until the animal achieved the criterion score of 80 correct trials in 100 across five consecutive sessions. The time limit for the behavioral response was set initially to 10 sec, and to 5 sec at final stages of training; withholding the response beyond that limit was scored as error. The response of the animal toward the central food well was scored as an error. There was no correction for errors, *e.g.* an animal after making an incorrect response to central well was not allowed to correct itself by the response to the side well in the same trial, nor in the following one.

Task II: M-DNMS (Manipulatory Variant of DNMS)

Four cats, two of them males, were trained on this task. The procedures were the same as those used for DMS task, except that now the novel object was always baited on the choice test, requiring the animal to learn the rule of delayed nonmatching-to-sample.

Let us turn now to locomotor versions of these recognition memory tasks.

Task III: L-DMS (Locomotor Variant of DMS)

Four cats, one of them a male, were trained on this task. A schematic representation of the Nencki-type testing room [Konorski, 1967] is shown in figure 3.

The cat cage (labeled as “St.c” in figure 3), measuring 45 cm x 45 cm x 50 cm was equipped with an opaque screen. To facilitate perception of the stimuli from the cat cage, only the stimulus array’s larger objects (15-20 cm high and 3-5 cm in diameter) were used in this and following tasks. The three stages of preliminary training were similar to those given in WGTA, except that the animals were shaped behaviorally to run to the food wells to retrieve rewards, and then to return to the cage and reenter it. The preliminary training was completed on average in 2 days. When the animal returned to the cage from the central food well after the sample presentation, the opaque screen was lowered for the delay period (5 s and 10 s in random order, as before), after which it was raised so that the animal could choose between the lateral food wells marked by sample and novel objects. On this task, as in Task I, the food

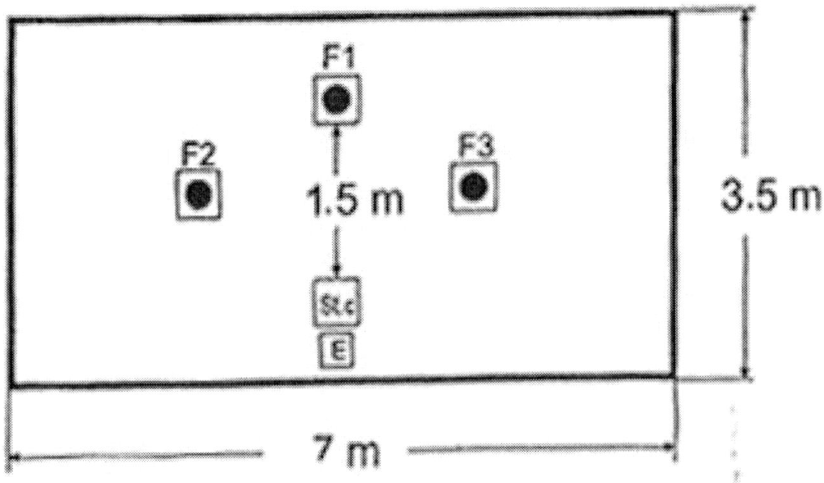

Figure 3. Schematic representation of the Nencki-type testing room used for Tasks III-V. Abbreviations: (St.c) start-cage; (F1, F2, F3) three identical food wells; (E) location of the experimenter. Arrow inside the figure (calibration bar) indicates the approximate distance between the start-cage and the food wells.

well marked by the sample object was always baited on the choice tests, requiring the animal to learn the rule of matching-to-sample. In all other respects also, the procedures for this task were the same as those used for Task I.

Both for the pseudotrials and during formal testing, objects were placed in front of the baited central food well (F1 in figure 3) for the sample presentation, and in front of the lateral wells (F2 and F3 in figure 3) for the choice test.

Task IV: L-DNMS (Locomotor Variant of DNMS)

Three cats, all males, were trained on task IV. All procedures were the same as those used in Task III, except that on the choice tests the novel object was always baited in the case of appropriate response, requiring the animal to learn the rule of nonmatching-to-sample.

Task V: L-DR (Locomotor Version of Spatial Delayed Response)

The four cats that had been trained on Task III were transferred to Task V. This task was adapted from the "place trials" of the object-location task described by Parkinson and coauthors [Parkinson, Murray, Mishkin, 1988], in which only spatial location was relevant for successful performance, object quality being irrelevant. For the sample presentation, a trial-unique object was placed in front of one of the three food wells, which were baited, and which the cat approached in order to retrieve the reward. For the choice test, the sample object appeared again in front of the same food well as before, and this food well was also baited as before, while an object identical to the sample appeared in front of one of the other two, unbaited, food wells. The spatial locations of the baited well and of the covered but unbaited well were selected pseudorandomly. In all other respects, this task was the same as Task III (L-DMS).

Tests with Extended Delays: Tasks I-IV

Immediately after completing training on the 10 s delay, each cat was retrained to criterion on the same task it has received before, but at a 30 s delay. Two weeks later each cat was tested for retention at the 10 s and 30 s delays, retrained to criterion if necessary, and then trained to criterion at both 5 min and 10 min delays. These tests with extended delays were presented in 20 trial daily sessions in pseudorandom order. Statistical assessments of the behavioral data were performed on the number of trials and errors preceding criterion at each training or retraining stage by Mann-Whitney U test [Hollander, Wolfe, 1973].

For the second subsection of this part only the M-DMS task was used for control and operated cats (bilateral lesions of the "rhinal region"); as general behavioral procedures of M-DMS testing were already outlined above, here we will consider only the surgical methods employed for this subsection. For this reason we performed another study of nonspatial visual recognition memory in new group of 12 adult cats of both genders (seven male, five female), weighing 3-3.5 kg, not previously used in experimental studies. Between experiments, animals were kept in the animal house in individual cages (1.5 x 1.0 x 1.0 m) in which they had free access to water. Food was given once daily, 20 h before testing. Experimental sessions were conducted 5 days per week. Same modified WGTA apparatus was used in this study as already described in preceding section of the Chapter. M-DMS task described previously was used for one-trial nonspatial visual recognition memory testing with all parametric data (trials per experimental day, criterion of performance, delay intervals, intertrial intervals and so on) just identical to previously described.

Surgery

After completion of preliminary training animals were divided into three independent groups each of four cats. In one group animals received lesions to the rhinal (i. e., perirhinal + entorhinal) cortical area (group R, n=4). In the next group, cats underwent all surgical manipulations except for lesioning of the structures of interest (sham-operated animals, SOP, n=4). The last group consisted of non-operated control animals (group C, n=4). All surgery was performed in strictly aseptic conditions on animals under deep Nembutal (40-60 mg/kg) anesthesia. Electrolytic lesions of cortical areas were produced by passing an anodal current of 2-5 mA through an electrode of stainless steel, diameter < 0.2 mm, insulated with epoxy resin throughout its length except the final 0.5 mm at the tip.

The coordinates for electrolytic rhinal lesions were selected using a stereotaxic atlas of the cat brain [Reinoso Suarez, 1961]. Cats in the SOP group underwent the same surgical procedures but no current was passed through the electrode. Coagulation of the whole rhinal region was produced by an average of 10-15 insertions of the electrode through small openings (diameter 0.5-1 mm) made in the cat's skull with a drill. Preliminary experiments yielded histological evidence showing that the diameter of the electrolytic lesion of cortical tissue produced by one insertion of the electrode and application of the current with the parameters noted above averaged 3-5 mm (figure 4 ,insert).

After all surgical procedures were completed, the margins of the surgical wound were sutured. The post-operative recovery period lasted 7-10 days, after which testing was started. The borders of the perirhinal and entorhinal cortex are shown in figure 5. By analogy with results obtained previously in other mammals [Krettek, Price, 1977; Witter, Groenewegen, Lopes da Silva, Lohman, 1989], the perirhinal region in cats is presumptively located along the posterior rhinal sulcus, occupying the base and lateral wall of this sulcus (figure 5, A,B).

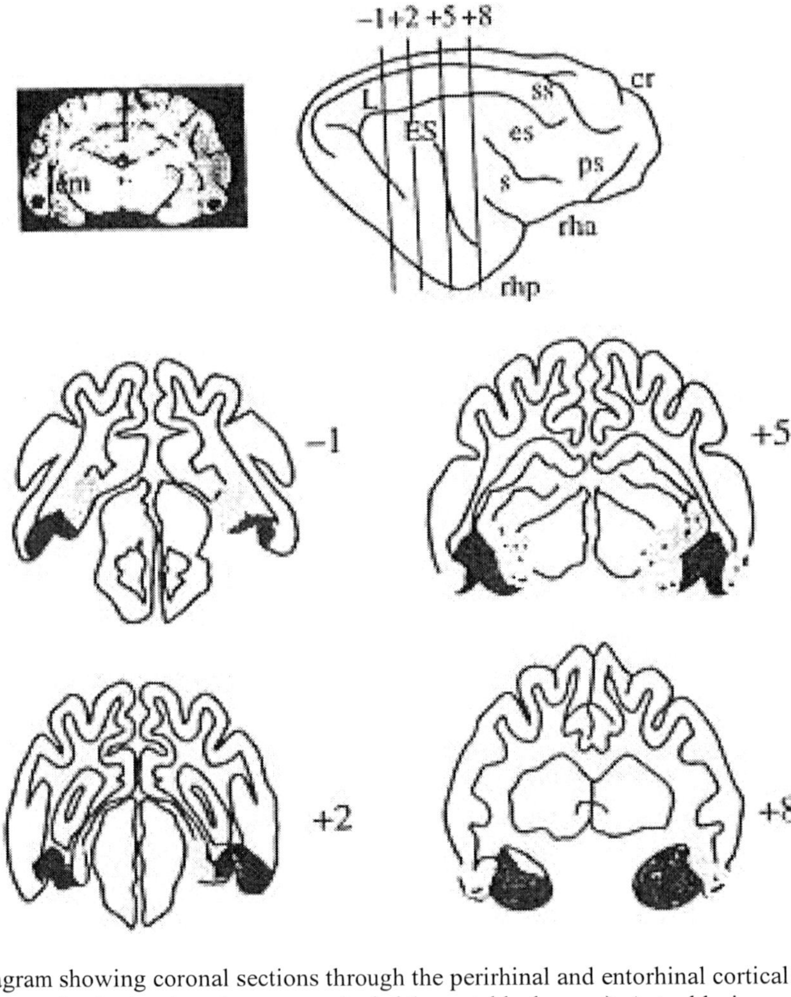

Figure 4. Diagram showing coronal sections through the perirhinal and entorhinal cortical areas of cats with brain lesions. Lesioning target areas are shaded (gray + black areas). Actual lesions are shown in black. The anteroposterior levels of the numbered sections (as per Reinoso-suarez stereotaxic atlas) are shown on the standard scheme of the lateral surface of the cat brain (above right). cr is the cruciatus, es the ectosylvian, L the lateral, ps the presylvian, rha the anterior rhinal, rhp the posterior rhinal, s the sylvian and ss the suprasylvian sulcus. The insert shows a transverse section at coronal level +5, showing the approximate size of the typical electrolytic lesion applied through a single electrode.

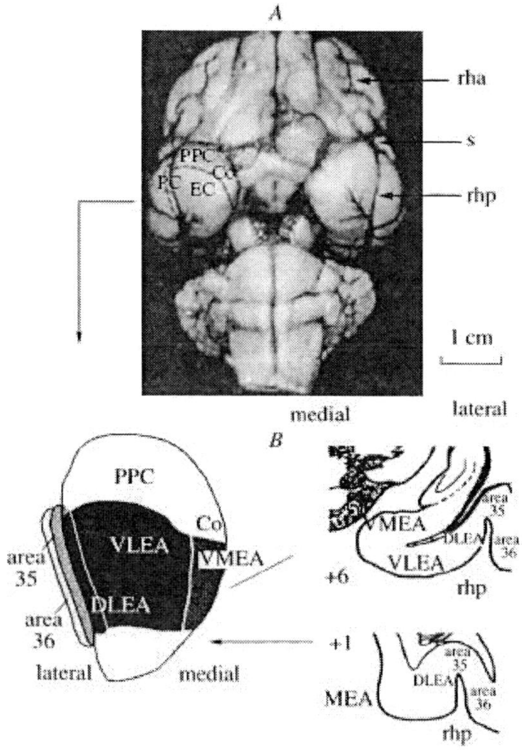

Figure 5. Photograph of the ventral view of a cat brain showing structures of the piriform cortex of the limbic lobe destined to be lesioned. (B) Enlarged diagram of the piriform cortex in the right hemisphere (left side of figure) shows cytoarchitectonic subdivisions of the perirhinal and entorhinal cortical areas. Arrows and diagrams indicate the levels of coronal sections through to be lesioned regions. Abbreviations: 35, 36, Brodmann's subdivisions of piriform cortex; Co, cortical amygdaloid nucleus; DLEA, dorsolateral entorhinal area; EC, entorhinal cortex; MEA, medial entorhinal area; PPC, prepiriform cortex; rha, anterior rhinal sulcus; rhp, posterior rhinal sulcus; s, sylvian sulcus; rs, VLEA, ventrolateral entorhinal area; VMEA, ventromedial entorhinal area.

Its anterior margin lies caudal to the sylvian sulcus (s), where the cortex in the depth of this sulcus is adjacent to the temporal cortex of the posterior composite sulcus. The posterior margin of the perirhinal area is located at the caudal margin of the posterior rhinal sulcus, where the latter splits into the medial-running pararecurrent sulcus [Kreiner, 1971]. The entorhinal cortex (EC) occupies the posterior part of the piriform cortex, which is the ventral part of the limbic lobe. The anterior margin of the entorhinal cortex is adjacent to the prepiriform cortex (PPC) and the cortical amygdaloid nucleus (Co). Laterally, the entorhinal cortex runs to the medial wall of the rhinal sulcus (rhp), where it is adjacent to field 35 of the perirhinal cortex, as medially it reaches the parasubicular area of the hippocampal formation. The posterior margin of the entorhinal area consists of the retrosplenial sulcus, separating the entorhinal cortex from the post- and retrosplenial cortical areas located on the medial surface of the hemisphere.

Table 1. Individual scores preceding criterion on manipulatory Tasks I and II (criterion sessions were not included into the trials and errors to criterion indices)

Task I: M-DMS			Task II: M-DNMS		
Cats	Trials	Errors	Cats	Trials	Errors
Alpha	190	66	Anna	180	91
Beta	260	92	George	200	75
Gamma	350	116	Strong	170	50
Delta	220	75	Nancy	220	84
	240	84	MEDIAN	190	80

Results and Discussion

In present study following results have been obtained.

Tasks I and II: M-DMS and M-DNMS

The data gathered on the manipulatory versions of delayed matching and nonmatching are presented in Table 1. The eight cats trained on one or the other of these two tasks learned in an average of 224 trials and 81 errors.

Although the group trained on DNMS learned slightly more quickly then the group trained on DMS, the difference was not significant. These results differ from those reported in monkeys [Mishkin, Delacour, 1975], which showed a distinct advantage in learning DNMS as compared with DMS. It was concluded from the latter set findings that monkeys have a preference for novelty, a preference that hastened their learning of the nonmatching rule. On the basis of the findings in the present study, it appears that cats either do not have this preference for novelty or do not exhibit it to the same degree as monkeys do.

These conclusions regarding a species difference in preference for novelty and the effects of this preference on learning in cats and monkeys are supported by a direct comparison between the scores of the two species. Comparing our cats with monkeys from Mishkin and Delacour [1975] study it must be noted that in these two cases different learning criteria were used (80% in 100 consecutive trials for cats vs. 90% in 40 consecutive trials for monkeys). From the statistical point of view such comparison at first may seem not to be meaningful. But let us consider the data from Mishkin and Delacour paper: it is evident that their monkeys in M-DMS task ("E-M" in their notation), before reaching 90% criterion, reached criterion of 80% correct responses within 40 trials (median for trials to reach this criterion - 320), after which their learning curve raised monotonously to higher levels of performance (90% correct responses in 40 trials); thus it may be assumed that their monkeys in criterial trials in M-DMS task performed at least as efficiently as our cats in their 100 criterial trials. As to the M-DNMS task ("E-NM" in Mishkin and Delacour notation) monkeys reached criterion of 80%

Table 2. Individual scores preceding criterion on locomotor Tasks III and IV (Criterion sessions were not included into the trials and errors to criterion indices)

L-DMS			L-DNMS		
Cats	Trials	Errors	Cats	Trials	Errors
Tetra	1060 (F)	520	#1	510	245
Shava	640	255	#2	725	350
Natsara	900	390	#3	1000 (F)	470
Tsotskhala	1040 (F)	435			

F denotes failure to achieve criterion within 1000 trials. Although two cats (Tetra and Tsotskhala) achieved criterion in additional 6o and 40 trials, respectively, while on the third cat (Natsara) training was disconnected after 1000 trials.

correct responses in at least 60 trials to criterion in spite of non-monotonous character of their learning curve [Mishkin Delacour, 1975; figure 1]. It is worthwhile to mention that according to our results cats after having achieved 80% correct criterion in 100 trials performed subsequent lengthened delay tasks at least at the same (80%) proficiency level in both tasks (Table IV). From these statements one may conclude that it is meaningful to compare our cats with monkeys from Mishkin and Delacour work mentioned above. The eight monkeys in the study by Mishkin and Delacour [1975] learned one or the other version of the two tasks in an average of 225 trials and 85 errors, an overall mean score remarkably similar to that of the eight cats in the present study. However, the monkeys' scores were distributed very differently between the two tasks: DMS, 360 trials and 150 errors; DNMS, 90 trials and 21 errors. As a result, although the cats learned the nonmatching rule significantly more slowly than the monkeys did (two-tailed Mann-Whitney $U_{4,4}=0$, P=0.03, for both trials and errors), they tended to learn the matching rule slightly more quickly than the monkeys, though the difference in this case is not significant.

Tasks III and IV: L-DMS and L-DNMS

The results obtained in the Nencki-type testing room are presented in Table 2.

Compared with the cats trained in the WGTA, those trained on the locomotor versions learned relatively slowly, with some animals even failing to reach criterion within the limit of 1 000 trials. The cats in this situation exhibited a strong tendency to return to the central food well on the choice test, suggesting that visuospatial strategies, which are well developed in cats [Warren, 1965; Warren, Warren, Akert, 1972], had interfered with their learning of the object matching and nonmatching rules. To examine how readily cats might learn a locomotor task in which memory for spatial cues was relevant and memory for object cues was irrelevant, we transferred the animals that had been trained on task III to Task V.

Task V: L-DR

As shown in Table III, all four cats, including the two that had previously failed to learn the locomotor version of delayed object matching, learned the locomotor version of spatial delayed response within just four sessions (i.e., 80 trials).

The results support the suggestion that the cats' strong tendency to use visuospatial strategies in the large testing room interfered with their learning of the object matching and nonmatching rules in that situation.

Performance at Extended Delays: Tasks I-IV

As shown in Table 4, those cats that had mastered object matching and nonmatching up to delays of 10 s committed about the same number of errors as before to master same task at a delay of 30 s.

Once they had attained the criterion score at 30 s, however, the cats re-attained this criterion almost immediately both when tested for retention after the two-week rest period and when delays were subsequently increased to 5 min and 10 min.

Table 3. Individual scores preceding criterion on the locomotor spatial memory task, Task V: L-DR (criterion sessions were not included into the trials and errors to criterion indices)

Cats	Trials	Errors
Tetra	60	25
Shava	40	15
Natsara	80	45
Tsotskhala	80	38

Table 4. Median errors to criterion for initial learning of Tasks I-IV at 10 s and 30 s, and for relearning after a two-week rest at both 10 s and 30 s and at extended delays of 5 min and 10 min (criterion sessions were not included into the trials and errors to criterion indices)

MEDIAN ERRORS						
Tasks Learning Relearning New learning						
Delays	10 s	30 s	10 s	30 s	5 min	10 min
M-DMS	84	90	0	2	6	2
M-DNMS	80	86	0	0	4	0
L-DMS	413	310	0	4	6	0
L-DNMS	350	325	0	0	4	0

Previous comparisons of the visual learning and memory abilities of cats and monkeys were based mainly on the use of the learning-set paradigm [Passingham, 1981; Warren, 1965], and these comparisons indicated inferior performance of cats on trial 2, a measure of one-trial memory. Based on the evidence gathered on the tasks used in the present experiment, however, the one-trial memory of cats seems to be quite comparable to that of monkeys, in terms of both overall speed of rule learning, at least in the WGTA, as well as level of retention over long intervals [Mishkin, 1982]. The results suggest that delayed matching- and nonmatching-to-sample are useful measures for investigating the neural basis of "long-term" visual memory in cats. During comparison of cats' performance on the visual recognition task with the performance of monkeys the following distinction is evident: cats perform the locomotor version of the recognition tasks poorly, due to their reliance on visuospatial strategies directed to find food in a particular place visited a bit earlier. Perhaps this distinguishes them, as a group, from monkeys, although this speculative point needs special comparative study. It might be supposed that this difference is caused to a substantial degree by cats' inherent tendencies to exhibit the predatory behavioral patterns directed mostly by visual checks of the spatial locations of relevant objects [Warren, 1965]. These tendencies, of course, should be present in primates too, but to a much lesser extent, because monkeys' manipulatory activity is more developed.

Altogether, it appears that cats, like monkeys, have a highly developed "long-term" visual recognition memory ability. At the same time our already reported results indicate the urgent need to investigate in cats the visuospatial recognition memory *per se,* although first of all, to be logically complete we heed now to report our data concerning possible brain mechanisms responsible to mediation of visual nonspatial memory in cats. Now, our following research step consisted in an attempt to uncover possible brain region sub serving nonspatial visual recognition in cats.

Morphological Monitoring

After behavioral testing, cats were overdosed with barbiturate, perfused via the heart with normal physiological saline followed by aldehyde fixatives. The brain was removed and photographed and cryomicrotome sections were cut; each fifth section was stained with thionine and was covered with a cover slip. Sections were examined histologically. The areas of cell loss and gliosis, along with the extent of the lesions, were superimposed on diagrams of intact coronal sections. Lesioned areas of structures were measured in square millimeters on drawings of each section. Lesion margins were determined using a simple digital system running on computer and subsequent manual calculation of lesion size. The volumes of lesioned tissue in the areas of interest were expressed as percentages of the normal volumes of the areas. The results for each case are presented in Table 5. Lesions to the rhinal area, regarded as combined lesions to the perirhinal and entorhinal areas, were incomplete (mean 75%, range 54-88% for the entorhinal cortex, mean 80%, range 64-89% for the perirhinal cortex). All animals showed some unplanned bilateral lesioning of the parasubiculum and

Table 5. Extent of damage to medial temporal lobe structures in percents to total volume of corresponding structures

Volume, mm^3	HF (406)			Para/Presub (70)			ERh (213)			PRh (233)			Compo (406)		
Cats	Lt	Rt	C	Lt	Rt	C	Lt	Rt	C	Lt	Rt	C	Lt	Rt	C
R. No. 1	0	2	1	61	51	56	98	83	91	56	72	64	6	3	5
R. No. 2	0	0	0	43	57	50	75	83	79	70	90	80	26	20	23
R. No. 3	14	12	13	32	34	33	95	44	70	90	88	89	9	21	25
R. No. 4	9	11	10	51	57	54	85	40	63	92	82	87	29	22	26
Mean	6	6	6	47	50	48	88	63	76	77	83	80	18	17	20

HF, Hippocampal formation; Para/Presub, parasubiculum plus presubiculum; ERh, entorhinal cortex; PRh, perirhinal cortex; Compo, area composita posterior according to Kreiner (1971); Lt, left; Rt, right;

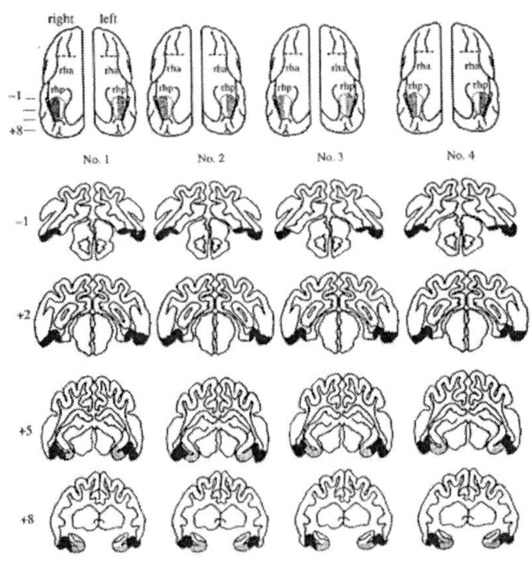

Figure 6. Diagram showing the basal surface of the cat brain (upper row); perirhinal lesions are shown in black, entorhinal lesions in dark grey. Below – reconstructions of perirhinal and entorhinal lesions (black) in four operated cats (R. No1, R. No 2, R. No. 3, and R. No. 4). Anteroposterior levels of numbered sections (as per Reinoso-Suarez stereotaxic atlas) are shown on the standard scheme of the ventral surface of the cat brain (upper diagram). Numbers indicate distances in millimeters from the interaural vertical plane. For further see captions of to figures 4 and 5.

presubiculum (mean 48%, range 33-56%; Table 5, figure 4), though unexpected lesioning of other areas outside the rhinal cortex was insignificant.Table 5. Extent of damage to medial temporal lobe structures in percents.

This included minor bilateral lesioning of the hippocampus itself (mean 6%, range 0-13%) and the area composita (mean 20%, range 5-26%) according to Kreiner's notation (1971), this being unilateral in the latter case – cat R. No. 4 (Table 5, figure 6).

Consequences of Lesioning of the Rhinal Area

Statistical analysis of the results of the behavioral experiments was performed using a computer running the statistical packages Statistica and SPSS 10 (Statsoft Incx., USA). Analysis of the data presented in Table 6 shows that the results obtained from performance of the "delayed matching-to-sample" test by intact and SOP cats could be combined into a single group, termed "control animals".

Single-factor ANOVA demonstrated a statistically significant intergroup difference between control and experimental cats with rhinal lesions to the brain ($F = 5.935$; $p = 0.009$). On the other hand, the same analysis showed that the post hoc Sheffe's test for multiple comparisons identified a significant difference between cats with lesions to the rhinal area and animals of the control group with a 10-sec delay ($p = 0.024$) but not with a 5-sec delay ($p = 0.999$). This is entirely consistent with results obtained from nonparametric analysis and, furthermore, is additional evidence for the absence of any influence of lesions to the rhinal area on the initial "delayed matching-to-sample" training.

It seems appropriate to start the discussion of the data obtained here from the point of view of Mishkin's hypothesis of the hierarchical organization of memory for one-trial perception of objects in primates [Mishkin, Suzuki, Gadian, Vargha-Khadem, 1997]. According to this hypothesis, recognition memory for different sensory modalities (including vision) is organized hierarchically: the hippocampal system receives its main inputs from widely disseminated areas of the neocortex via two main projections - one from the perirhinal and one from the parahippocampal cortex. The first of these projections (the "ventral stream")

Table 6. Individual scores preceding criterion (criterion sessions were not included into the errors to criterion indices) in DMS Task tested under "short" (5 sec, 10 sec) delay conditions in normal, sham operated (control animals groups) and operated rhinal cats. (criterion – no more than 20 errors in 100 consecutive trials

GROUPS:	ANIMALS	DELAYS	
		5 sec	10 sec
NORMAL CONTROL(n=4)	# 1	66	55
	# 2	92	100
	# 3	116	90
	# 4	75	48
SHAM OPERATED CONTROL (n=4)	# 1	102	62
	# 2	69	70
	# 3	75	102
	# 4	100	65
RHINAL OPERATED EXPERIMENTAL (n=4)	# 1	90	120
	# 2	87	110
	# 3	100	130
	# 4	105	150

arises from different sensory modalities important for analysis of stimulus quality, while the latter arises from the neocortical areas important for analysis of stimulus locations (the "dorsal stream"). We will not discuss dorsal stream here in this section as, according to the hypothesis, it is responsible for spatial memory, which was not addressed in the present part of this Chapter. The ventral stream, which is important for the organization of visual nonspatial memory, consists of "higher visual areas" (the inferior temporal cortex) and the perirhinal and entorhinal cortical areas; information is projected from these to the hippocampuses.

This raises the important question: is the ventral stream (or some analog!) present in carnivores? Verification of the existence of such a stream in cats clearly requires knowledge as to whether cats generally have an albeit functional analog (according to the results of Hodos and Campbell [1969] ideology) of the "inferior temporal cortex" in primates. Unfortunately, there are very few data on this important question and these data are contradictory: some investigators have shown the existence of such an analog in cats [Campbell, Jr., 1978; Hara, Cornwell, Warren, Webster, 1974; Natishvili, 1979], while others refute its existence [Gross, 1973]. In general, further investigations are needed to obtain a more strongly grounded interpretation of our data from this point of view. Our data suggest the conclusion that visual recognition in cats is mediated by the same cortical area as in monkeys. It should, however, be emphasized that our conclusion, like the analogous conclusion of American investigators with respect to visual recognition in primates [Murray, Bachevalier, Mishkin, 1989], is restricted only by short-term recognition memory. In our studies, animals of the experimental group (with lesions to the rhinal area) did not differ from controls on testing of recognition with delays shorter then 10 sec. This is evidently further support for the view that visual perception of an object and memory of it (even in the short-term version) depends on different brain structures in mammals. Overall, an important conclusion follows from the present data: visual recognition memory can be regarded as an "evolutionary invariant" in different mammal species. However these species have different ethological and ecological characteristics: it is sufficient to say that monkeys are diurnal, while cats are nocturnal. To summarize obtained results we can make following main inferences: cats trained in the our modified WGTA learned the two tasks at about the same rate, on average, as that reported for monkeys. However, unlike monkeys, whose strong preference for novelty facilitates their learning of the nonmatching rule and retards their learning of the matching rule, the cats learned the two different rules at about the same rate, suggesting that cats do not share the monkey's strong preference for novelty. Iп contrast to their relatively rapid learning of the manipulatory versions of the two tasks, cats learned the locomotor versions only slowly or even failed to learn. Experimental analysis indicated that a major source of the cats' difficulty on these locomotor versions was interference from a strong tendency in the large testing room to use visuospatial strategies. Thus let us turn now to the study of visuospatial recognition memory in cats.

In: Cats
Editors: T. Zakrevsky and V. Lachinov
ISBN 978-1-62081-043-9

Chapter VII

BEHAVIOURAL ECOLOGY OF THE EUROPEAN WILDCAT (*FELIS S. SILVESTRIS* SCHREBER 1777), DOMESTIC CAT (*FELIS S. CATUS* L.) AND THEIR HYBRIDS: CURRENT IMPLICATIONS IN WILDCAT CONSERVATION

***Estelle Germain*[*,1] *and Marie-Lazarine Poulle*[*,*,2]**
[1]CROC, Carnivores Recherche Observation Communication, Fonteny, France
[2]Université de Reims Champagne Ardenne, Centre de Recherche et de Formation en Eco-éthologie (URCA-CERFE), France & UFR Médecine, Laboratoire de Parasitologie-Mycologie, Reims, France

WHAT DOES THE EUROPEAN WILDCAT LOOK LIKE?

The European wildcat (*Felis s. silvestris*) is one of the last small wild felines in Europe. Contrary to popular belief, it is not larger or heavier than most domestic cats (*Felis s. catus* also named *Felis catus*). Its weight has often been exaggerated in older publications due to confusions between pounds and kilograms or the voluntary omission of light specimens in analyses (Condé & Schauenberg 1971). The average weight of the European wildcat is around 5 kg for males and 3.5 kg for females (Condé & Schauenberg 1971) even if the

* estelle.germain@croc-asso.org.
** Université de Reims Champagne Ardenne, Centre de Recherche et de Formation en Eco-éthologie (URCA-CERFE), 08240 Boult-aux-Bois, France & UFR Médecine, IFR 53, Laboratoire de Parasitologie-Mycologie, EA3800, 51 rue Cognacq-Jay, 51096 Reims, France; marielazarine.poulle@cerfe.com.

European wildcat seems bulkier and more robust than most domestic cats. The head-body length varies between 52 and 65 cm for males and 48.5 and 57 cm for females, and tail length varies between 26 and 34.5 cm for males and 25 and 32 cm for females (Schauenberg 1977a).

The wildcat's coat is not spotted, with few stripes and presents three main characteristics: a single, well distinguished black band from mid-back to the base of the tail; a thick, ringed tail (two or three very visible rings) with a black tip; and black stripes from the forehead region up to the base of the neck (Stahl & Léger 1992; Léger et al. 2008). The winter coat is darker and more silver than the summer coat and the underside varies in individuals. Three hues can be distinguished: grey, light beige and sometimes black (Stahl & Léger 1992; Jackson et al. 1996). The chin and mouth are off white, the nose is pink ringed with black and the whiskers are white. Stripes appear on the flank but are much less marked than those of a tabby domestic cat and they are not connected on the backbone region and disappear on its underside (Stahl & Léger 1992; Jackson et al. 1996).

The European Wildcat: A Threatened Felid

European wildcat populations were broadly distributed from the Holocene epoch through the Middle Ages (Sommer & Benecke 2006) but during the 18th and 19th centuries the number and size of wildcat populations drastically decreased everywhere in Europe due to trapping for fur and poisoning (Stahl & Léger 1992; de Planhol 2004). This decline due to pressure exerted by humans was then accentuated by the parallel loss and fragmentation of wildcat habitats (Stahl & Léger 1992; Sunquist & Sunquist 2002). At the beginning of the 20th century, European wildcat populations were close to extinction (McOrist & Kitchener 1994; Stahl & Artois 1994; Sunquist & Sunquist 2002). Nowadays, the European wildcat is distributed in fragmented populations ranging from Russia to Portugal and from Scotland to the Near East (Stahl & Artois 1994; Nowell & Jackson 1996; Sunquist & Sunquist 2002). Its geographical distribution covers a large part of Europe, but is broken up into separate geographical zones that are often quite far from each other. It covers the Pyrenees and Iberian peninsula, the Apennine Mountains and southern Italy, the Balkans and Carpathian Mountains, Asia minor, Caucasus, Palatinate, Harz and Eifel in Germany, north-eastern France, Swiss Jura, Luxemburg, the Belgian Ardennes and northern Scotland (see Stahl & Artois 1994 for a review; Léger et al. 2008).

Since the 1970s, the European wildcat has been considered to be an endangered species and is strictly protected under Appendix II of the Bern convention (Council of Europe 1979) and Appendix IV of the European Habitats Directive 92/43/EEC (Council of Europe 1992). Given its wide geographical distribution, it is listed as 'Least Concern' on the Red List of the International Union for Conservation of Nature (Driscoll & Nowell 2010) but it has nevertheless been mentioned as a protected species in Appendix 2 of the Convention on International Trade in Endangered Species (see UNEP-WCMC 2012). In France, the European wildcat has been protected since the implementation of the laws relative to nature protection in 1976 (Journal Officiel de la République Française 1976). Despite its full legal protection in most European countries and an apparent return of populations in certain regions in Europe (Stahl & Artois 1994; Léger et al. 2008; Klar et al. 2008; Say et al. 2011), the

European wildcat is still considered to be seriously threatened by trapping, hunting, poisoning (use of toxic chemicals in agriculture or in poisoning of wildlife), road accidents or illegal direct killing (Stahl & Artois 1994; McOrist & Kitchener 1994). Numerous other threats are still unidentified due to a lack of knowledge about European wildcat ecology (Lozano et al. 2007).

One of the main threats highlighted by different authors is the reduction and alteration of forests due to the development of farming and extensive road networks as well as poorly adapted forestry techniques (Hubbard et al. 1992; Stahl & Artois 1994; McOrist & Kitchener 1994; Oliveira et al. 2007, 2008; Léger et al. 2008; Klar et al. 2009; Hertwig et al. 2009; Krüger et al. 2009). The loss of habitat induces other threats like ecological niche loss and isolation of wildcat populations, and thus increases the risk of genetic diversity loss and the risk of interactions with domestic cats (Hertwig et al. 2009; O'Brien et al. 2009; Krüger et al. 2009). Interaction with domestic cats may be deleterious for the European wildcat in two ways: diseases transmission and hybridization (Randi et al. 2001; Stahl & Artois 1994; McOrist & Kitchener 1994).

General Aspects of Hybridization

Hybridization is the result of reproduction between individuals from two distinct populations characterized by their genetic proximity whatever their taxonomic status (Arnold 1992; Rhymer & Simberloff 1996; Seehausen 2004). The resulting offspring are first generation hybrids (F1), viable or not, fertile or sterile. If these hybrids are viable and fertile, they can reproduce together to produce the second through nth generations (F2…Fn) or they can reproduce with individuals from parental populations being then responsible for introgression (Arnold 1992; Seehausen 2004). This phenomenon is known to occur in the Carnivora order and notably in the Felid family (see Wayne & Brown 2001 for a review). In this way, the European wildcat can breed with the domestic cat to produce viable and fertile hybrids (Ragni 1993) and these hybrids may be responsible of the introgression of domestic cats' genes into wildcat populations (Beaumont et al. 2001; Hertwig et al. 2009).

Hybrids do not necessarily have intermediate phenotypes corresponding to a mixture of the parental phenotypes and, reciprocally, all intermediary forms of phenotypes are not necessarily indicative of hybrids (Dowling & Secor 1997). The difficulty in differentiating hybrids morphologically from parents probably partially explains why hybridization was long considered to be a rare phenomenon in animals (Arnold 1997; Dowling & Secor 1997; Allendorf et al. 2001). According to Mallet (2005), only 10% of animal species undergo hybridization and introgression mechanisms.

Hybridization generally occurs when two populations are found in sympatry after the breakdown of a physical barrier that had isolated them geographically until this event (Bullini 1994; Seehausen 2004). For example, the expansion of forests during glacier periods permitted previously isolated populations of bears to expand their geographic ranges and join together naturally. As a result, eastern and western black bear (*Ursus americanus*) populations hybridized together in Canada (Wooding & Ward 1997) as well as eastern and western European brown bear (*Ursus arctos*) populations in Europe (Kohn et al. 1995). In both cases, hybridization was conducive to the formation of a hybrid zone respectively located in the western Rocky Moutains for black bears (Wooding & Ward 1997) and in Romania for brown bears (Kohn et al. 1995). Hybridization between populations can occur in natural conditions (i.e. 'natural hybridization') but it increases in frequency with environmental modifications due to human activity (i.e. 'anthropogenic hybridization'; Rhymer & Simberloff 1996; Allendorf et al. 2001).

Hybridization can have harmful consequences in terms of conservation since producing a hybrid offspring can compromise the persistence of rare species (Rhymer & Simberloff 1996; Dowling & Secor 1997; Allendorf et al. 2001). This problem can occur within a short time frame: extinctions of species endangered by hybridization can sometimes occur in only five generations (Wolf et al. 2001). Hybridization can therefore threaten the genetic integrity of certain species whose genotype could disappear little by little (further to the introgression of genes from another species) to the benefit of the best represented species in terms of numbers of reproductive individuals (Rhymer & Simberloff 1996). The regression of certain low-density populations of the European mink (*Mustela lutreola*) in Europe could therefore be linked to hybridization with the European polecat (*M. putorius*) (Maran & Henttonen 1995; Cabria et al. 2011). Hybridization and introgression are also currently considered to be a major threat to European wildcat populations, compromising their genetic integrity (Randi et al. 2001; Pierpaoli et al. 2003; Randi 2008; O'Brien et al. 2009; Driscoll & Nowell 2010; Say et al. 2011).

Genetic and Geographical Proximity between European Wildcats and Domestic Cats

The classification of Felids is very complex and is always a subject of discussion (see introduction of Kitchener & Rees 2009). However, older classifications, founded on morphological criteria, and recent classifications, benefiting from molecular genetic tools, generally agree that the European wildcat belongs to the polytypic species *Felis silvestris* (Ragni & Randi 1986; Randi & Ragni 1991; Ragni & Possenti 1996; Randi et al. 2001). Recently, Driscoll et al. (2007) concluded that the European wildcat, the African wildcat (*Felis s. lybica*), the Asian wildcat (*Felis s. ornata*) and the domestic cat are distinct sub-species resulting from the same polytypic species. The European wildcat line may have diverged between 173 000 and 230 000 years before present (B.P.).

The domestication of the African wildcat, that is a small rodent predator, has been dated several times (see Sunquist & Sunquist 2002 and O'Brien et al. 2008). It was thought for a long time that the process was the result of Egyptian civilisation (see Kitchener 1991; Jackson et al. 1996; Sunquist & Sunquist 2002). However, Vigne et al. (2004) found cat bone remains in a human burial site in Cyprus and they thus assume that taming of African wildcats may have begun 9500 to 9200 years B.P. with the arrival of the first Neolithic settlers on the island. The installation of wildcats in villages could have been favoured in order to control the mouse population and protect crops and cereal stocks from loss and damage due to rodents (Sunquist & Sunquist 2002). The Greeks introduced the domestic cat, resulting from this domestication process, in Italy during the 5th century B.C. (Serpell 1988). It was spread throughout Europe with Roman expansions during which armies certainly considered them as "a camp follower and companion" and became abundant in European continent in the $4^{th} - 5^{th}$ centuries (Serpell 1988; Kitchener 1991; Sunquist & Sunquist 2002). The domestic cat is currently present on all continents in rather varied habitats (see Turner & Bateson 1988) and was introduced onto most islands, where its populations developed rapidly, including in extreme conditions such as those seen in the Kerguelen Islands in Antarctica (e.g. Say et al. 2002).

Whether or not it was voluntary, the introduction of the domesticated form of the African wildcat on the European and Asian continent broke the geographical isolation that up until then had preserved wildcat populations from hybridization. Currently, like in European wildcat populations, those from South Africa and the Near East may be threatened by hybridization with the domestic cat (Nowell & Jackson 1996). It is possible that hybridization exists since the European wildcat and domestic cat have begun living in sympatry, i.e. about 2000 years ago (Balharry et al. 1994, 1997; Daniels & Corbett 2003) though it may have been rare or inexistent up until the beginning of the 20th century (Stahl & Léger 1992). Currently, the frequency of contact between wildcats and domestic cats may have increased because of profound environmental modifications due to humans. These modifications may have led to an imbalance between the numbers in populations of wildcats and domestic cats (Stahl & Léger 1992; McOrist & Kitchener 1994). This imbalance, associated with a fragmented distribution of wildcats, and along with the fact that at the beginning of the 20th century wildcats began recolonizing former habitats that then became occupied by domestic cats (Hubbard et al. 1992), may have brought about an increase in hybridization frequency in certain European wildcat populations.

Hybridization Between the Wildcat and Domestic Cat: A New Phenomenon?

It is however impossible to know when hybridization began in wildcat populations since there is no European wildcat population that remained isolated from domestic cats and that could thus serve as a reference population, from either a genetic or morphological point of view (French et al. 1988; Daniels et al. 1998, 2001; Randi et al. 2001; Daniels & Corbett 2003). The characteristics of the wildcat before the introduction of domestic cats in Europe are unknown and have necessarily evolved since the introgression of domestic cat genes in wildcat populations may have occurred (Yamaguchi et al. 2004; Kitchener et al. 2005). According to Balharry & Daniels (1998), the typical specimen described by Schreber in 1777 is probably not very representative of the original ancestor of the wildcat, as it was already the result of hundreds of years of sympatry with the domestic cat. All the naturalized collections of wildcats in Scottish museums were designed based on this typical specimen, rejecting or systematically relegating any individual to the category of hybrid if its characteristics differed too much from it (Balharry & Daniels 1998). The result is a collection of cats that look alike and giving the impression that there is a distinct population that could be identified based on morphological criteria, such as coat (Balharry & Daniels 1998). This approach did not take into account the morphological and anatomical variability in characteristics of wildcats highlighted by French et al. (1988), Ragni & Possenti (1996) or Daniels et al. (1998).

Cats with 'European wildcat characteristics' could be wildcats or hybrids and, conversely, cats with 'dubious European wildcat characteristics' could be wildcats (Krüger et al. 2009; Ruette et al. 2011). Thus, in a context where hybridization occurs, it does not seem realistic to distinguish wildcats from hybrids by using the morphological, morphometric or anatomical criteria proposed to distinguish wildcats from domestic cats (see examples of these criteria in Suminski 1962; Schauenberg 1969, 1977b; Ragni & Randi 1986; Ragni & Possenti 1996; French et al. 1988; Daniels et al. 1998). The difficulty of distinguishing 'pure' wildcats from hybrids and defining specific diagnostic characteristics is even more marked given the different degrees of hybridization observed in Europe from one population to another (Yamaguchi et al. 2004; Kitchener et al. 2005; Platz et al. 2011). These diagnostic characteristics should thus be population-specific and should be defined for each region/population where differences in the degree of hybridization are encountered (Yamaguchi et al. 2004; Platz et al. 2011). However, diagnostic characteristics alone are not sufficient to ensure unambiguous hybrid identification (Hertwig et al. 2009; Krüger et al. 2009; Ruette et al. 2011). As the identification of hybrids is related to the validity of the identification of parental taxa (Krüger et al. 2009), using molecular biology techniques is of high interest and more precise for identifying wildcats, domestic cats and hybrids when morphological and anatomical identification is uncertain (Oliveira et al. 2007; O'Brien et al. 2009). However, these methods did not exist even a few decades ago.

As a consequence, it is not always possible to know if the apparent preservation or even expansion of range areas reported for certain populations of wildcats (e.g. Parent 1975; Corbett 1978; Dötterer & Bernhart 1996; Heltai et al. 2006) is a sign of 'good health' of the populations or if they result from an increase in hybridization frequency (Léger et al. 2008). The expansion of wildcat populations in different regions of Europe between 1930 and 1960 (especially after the Second World War - Haltenorth 1957) could partly be attributed to hybridization (Suminski 1962; French et al. 1988; Stahl & Léger 1992). This is probably the case in Italy where hybrids have been found on the edges of the wildcat distribution (Randi et al. 2001; Lecis et al. 2006) and in northern Scotland (Beaumont et al. 2001). However, this does not appear to be the case in France since hybrids are not more frequent in emergent or re-emergent zones of European wildcats than in areas in which wildcats have been historically present (Say et al. 2011).

Currently, a high degree of hybridization has been reported in wildcat populations in Scotland (Daniels et al. 2001; Beaumont et al. 2001) and Hungary (Pierpaoli et al. 2003; Lecis et al. 2006), whereas hybridization appears to be rare in Italy (Randi et al. 2001; Pierpaoli et al. 2003; Lecis et al. 2006), Portugal and Spain (Pierpaoli et al. 2003; Oliveira et al. 2007, 2008), Germany (Eckert & Hartl 2005; Eckert et al. 2010), France (O'Brien et al. 2009) and Bulgaria (Pierpaoli et al. 2003). However, the frequency of the emergence of hybrids can also differ between populations within a country, as is the case in Germany (see Hertwig et al. 2009; Krüger et al. 2009). The differences observed from one population to another (between countries or within a country) could be the result of a combination of historical, demographic and ecological factors, such as low numbers of wildcats and

proximity with high density domestic cat populations, habitat and landscape structure, urbanization or scarce availability of preys (Stahl & Artois 1994; Pierpaoli et al. 2003; Lecis et al. 2006; Oliveira et al. 2008; Randi 2008). The
relative importance and the nature of these factors that act as isolating barriers still remains to be determined (Stahl & Artois 1994; Pierpaoli et al. 2003; Lecis et al. 2006; Oliveira et al. 2008; Hertwig et al. 2009).

Factors that Promote Hybridization

The intensification of mechanization, industrialization, the development of agriculture, demographic growth and urbanization on-going for a century now, have resulted in, among other things, directly influencing the geographical distribution of numerous animal populations, particularly by reducing their populations' numbers (Paterson 1999; Dudley et al. 2002; Gaston 2003), with some being concerned by hybridization. In vertebrates, hybridization generally occurs in a situation of local rarity of reproductive partners for at least one of the two taxa, often due to modifications in habitats or the introduction of non-indigenous species (Rhymer & Simberloff 1996; Allendorf et al. 2001). The classic case (Figure 1) is where, after the disappearance of a physical barrier, a native population (population A) is in sympatry with a genetically close population but in higher numbers (population B). Population A is then weakened in terms of reproducing individuals and gene flow. The production of a fertile hybrid offspring (population C) is thus problematic in this case since the rare species is in a situation where it is forced to reproduce with the other species and its own hybrids. It is, in particular, the situation faced by the Canada lynx (*Lynx canadensis*) against the Bobcat (*L. rufus*) (Schwartz et al. 2004). Long term, population A could disappear to the benefit of a mixed population made up of population B and a population of hybrids.

This numerical imbalance between populations - that can favour hybridization - is generally the result of two categories of factors: those that weaken population A and those that allow population B to have a high population dynamic. Population A could be weakened and/or in lower numbers than population B (1) if it is at the edge of its range area (Arnold 1997; Seehausen 2004), such as the European wildcat in Italy (Lecis et al. 2006), (2) if its habitat is of poor quality and/or fragmented, such as the Northern Spotted Owl (*Strix occidentalis caurina*) threatened by the destruction of its habitat, that hybridize with the Barred Owl (*S. varia*), whose distribution is progressively expanding to the Northwest of the Pacific (Peterson & Robins 2003; Kelly & Forsman 2004), (3) if it is subject to high hunting and trapping pressures (see examples of disappearance of species under human pressure in de Planhol 2004), or (4) if it undergoes high mortality pressure due to competitors, disease or predation (see notably Wayne & Brown 2001). Inversely, population B could be extremely dense since individuals benefit from care and food sources (domestic populations) or due to a reduction in mortality or competition pressure. As an example, in North America, drastic reductions in Gray Wolf (*Canis lupus*) populations have allowed the Coyote (*Canis latrans*)

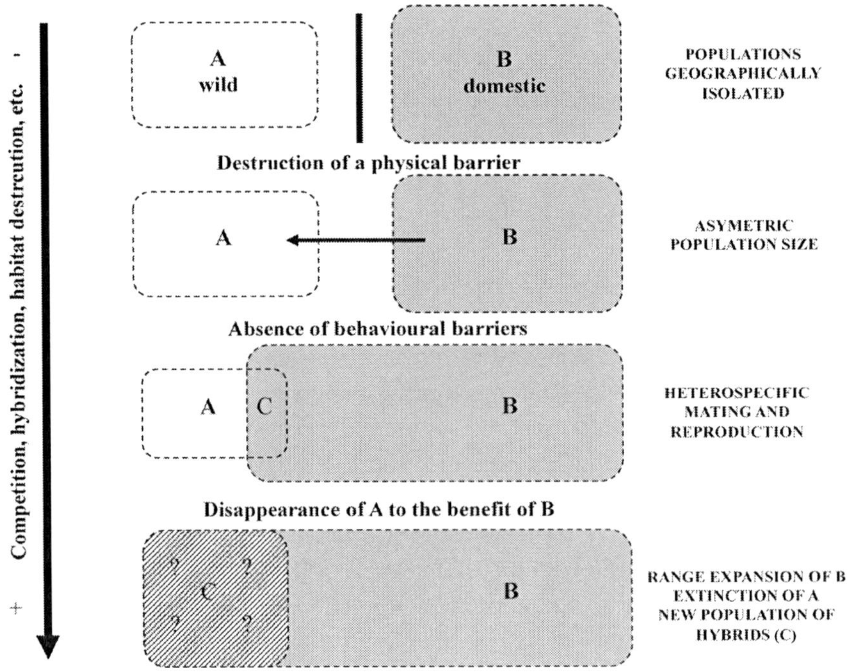

Figure 1. A classic case of hybridization in animal populations.

to increase its own range areas, which has led coyotes to hybridize with gray wolves, whose remaining populations were in numerical imbalance (Lehman et al. 1991; Moore & Parker 1992).

The European wildcat's situation sums up this combination of factors that could favour hybridization: the number of most of its populations has decreased during the 19th and the 20th century (Stahl & Léger 1992). In parallel, domestic cat populations, commensal with humans, have increased in number after the development of agriculture (Turner & Bateson 1988; Vigne & Guilaine 2004; Driscoll et al. 2007). In populations of wildcats with limited numbers, domestic cats may have been the closest and most easily accessible sexual partners than those of the same taxon (Suminski 1962).

In the Iberian Peninsula, hybridization is thought to be consecutive to the loss and fragmentation of wildcat habitat, the progressive expansion of urban zones and the rarity of prey, which favoured reproductive interactions with domestic cats (Oliveira et al. 2007). In France, the population of wildcats seems to have lived a "long history of hybridization" in the 20th century (O'Brien et al. 2009), certainly due to the low density and fragmentation of wildcat populations, whereas, on the contrary, populations of domestic cats have spread and extended their distribution (Stahl & Léger 1992; Léger et al. 2008). This could have also been the case for the evolution of wildcat populations in Britain (Hubbard et al. 1992).

However, if the demographic and ecological circumstances that favour hybridization in wildcat populations have begun to be identified, they are poorly understood and the ecological and behavioural processes involved in hybridization are still poorly studied.

Behavioural Ecology: Left Out in Studies on Hybridization Causes

The vast majority of studies on carnivores involved in hybridization have been genetic studies that lead to hypotheses on potential hybridization scenarios (e.g. Lehman et al. 1991; Maran & Henttonen 1995; Davison et al. 1999; Vilà & Wayne 1999; Beaumont et al. 2001; Verardi et al. 2006; Hailer & Leonard 2008; Cabria et al. 2011). Very few studies based on behavioural ecology have been carried to study hybridization in carnivores, by focusing simultaneously on the parental taxa involved and on their hybrids. In the case of the European wildcat, hybridization has also mainly been approached from a strictly genetic point of view (e.g. Daniels et al. 2001; Beaumont et al. 2001; Randi et al. 2001; Pierpaoli et al. 2003; Lecis et al. 2006; Oliveira et al. 2007, 2008; O'Brien et al. 2009; Eckert et al. 2010). As a result, several authors pointed out that the ecological, biological and behavioural factors influencing hybridization or introgression have to be identified (Kitchener et al. 2005; Oliveira et al. 2008; Randi 2008). There is notably a need for studies on wildcat home-ranges occupation by domestic cats (Sarmento et al. 2009), on the respective ecological niches of wildcats, domestic cats and hybrids (O'Brien et al. 2009) and on the ecology of domestic cats and hybrids to formulate current guidelines for European wildcat conservation (Lecis et al. 2006). Many questions remain about the conditions in which the matings between wildcats and domestic cats occurs, and especially about the terms of coexistence between them.

When geographical barriers that isolated populations that could hybridize are broken down, it is the behaviour and the ecology of individuals that influence the probability of meeting and therefore the possibility of mating (see Arnold 1997; Orr & Smith 1998; Mallet 2005; Schwenk et al. 2008). The probability of meeting between individuals can thus be limited by the fact that individuals live in distinct habitats (ecological isolation), reproduce at different times (temporal isolation) or have different pre-pairing behaviours (behavioural isolation). More specifically, when the habitat of one of the populations concerned is fragmented into small areas, social dysfunctions can appear in animals, which often need to be in proximity of several individuals (and thus in continuous areas of the habitat) to improve their chances of reproduction (e.g. Jaeger et al. 2005). This can push individuals to move over wide areas in their search for a reproductive partner, which increases the risk of meeting individuals from another population. Thus, to identify the circumstances of the appearance of hybrids in populations, the movement patterns of individuals as well as the range of their movements should be taken into account. For example, Carr et al. (1986) advise taking into account the specific dispersion pattern of each sex to understand why hybridization occurs between the Virginia deer (*Odocoileus virginianus*) and the Mule deer (*O. hemionus*).

Except for Biró et al. (2005) and Germain et al. (2008, 2009) studies, the ecological and behavioural aspects of the biology of wildcats, domestic cats and their hybrids in a context where hybridization has been identified are poorly explored. According to Corbett (1979), wildcats may exhibit very agonistic behaviour towards domestic cats. This behaviour may limit the entry of domestic cats into wildcat home-ranges (Hubbard et al. 1992) and thus, act as a barrier to hybridization (Easterbee 1991 *in* Hubbard et al. 1992). However, it is now known that hybridization occurs. The opportunities of meetings between wildcats, hybrids and domestic cats - that influence the risks of hybridization and introgression – may depend in large part on their respective ecological niches (O'Brien et al. 2009) and for instance, on the importance of the likeness between their food habits. Data on the degree of overlap between their diets could thus provide useful information on their chances of meeting, while data on the type of food items they share could provide information on the type of habitat where the meetings may occur. Furthermore, data on the temporal and space use patterns of wildcats and domestic cats in relation to their mating behaviour is needed in order to identify the behavioural processes that may favour or act against interbreeding.

Studies have been carried out on the behavioural ecology of European wildcats (e.g. Corbett 1979; Stahl 1986; Liberek 1999), particularly on their diet (e.g. Condé et al. 1972; Aymerich 1982; Riols 1988; Sarmento 1996; Tryjanowski et al. 2002; Moleón & Gil-Sánchez 2003; Malo et al. 2004; Biró et al. 2005; Lozano et al. 2006; Germain et al. 2009; Piñeiro & Barja 2011), on their space use patterns (e.g. Artois 1985; Stahl et al. 1988; Wittmer 2001; Biró et al. 2004; Sarmento et al. 2006; Germain et al. 2008; Klar et al. 2009; Monterroso et al. 2009), on the characterization of their habitat and its use (e.g. Wittmer 2001; Mermod & Liberek 2002; Lozano et al. 2003; Sarmento et al. 2006; Klar et al. 2008; Jerosch et al. 2010) and on their reproduction activity (e.g. Condé & Schauenberg 1969, 1974; Stahl et al. 1984). Studies have also been carried out on the behavioural ecology of domestic cats in rural environments (Hansen 2010; Silva-Rodriguez & Sieving 2011; Tschanz et al. 2011), but only the studies by Corbett (1979), Biró et al. (2004, 2005) and, recently, those by Sarmento et al. (2009) have focused on the ecological and behavioural aspects of relationships between wildcats and domestic cats in sympatric populations studied simultaneously. And, to date, only the studies by Biró et al. (2005) and our own studies (Germain et al. 2008, 2009) have given information on the behaviour and ecology of hybrids.

A Study of the Behavioural Ecology of Wildcats, Domestic Cats and their Hybrids in France

We sought to identify the differences and similarities between wildcats, domestic cats and hybrids by focusing on the way they exploit their food resources and share space, and we focused our attention more specifically on the behaviour of hybrids, which is not well known. The reader can refer to Germain et al. (2008, 2009) to have more specific information on the methodology that is quickly presented in this chapter. Our study was carried out in France, a country that is home to one of the highest populations of European wildcats (Léger et al. 2008; O'Brien et al. 2009; Say et al. 2011; Ruette et al. 2011). Wildcats are present in two distinct and unconnected areas. The main area of wildcat presence is located in the north-eastern quarter of France (Champagne-Ardenne, Bourgogne, Lorraine, Franche-Comté and Alsace) and has been expanding in the adjacent regions since the 1980s (see Léger et al. 2008; Say et al. 2011). It is continuous with wildcat populations in Germany, Belgium, Switzerland and Luxemburg (Léger et al. 2008) but is structured into two genetically distinct populations (Say et al. 2011). The second French wildcat area is located in the southwest of France, in the Pyrenees (Léger et al. 2008; Say et al. 2011). The presence of hybrids is known in France (O'Brien et al. 2009) and not limited to the edge of the wildcat distribution but spread over the entire range area of European wildcats (Say et al. 2011).

We assessed food habits of wildcats, domestic cats and their hybrids through the identification of undigested remains found in the stomachs of 59 road-killed adult cats (Germain et al. 2009). French Hunting and Wildlife Office (ONCFS) rangers collected cats from 2001 to 2006 at any time of the year and in the entire French wildcat range area. These cats were genetically identified by O'Brien et al. (2009). The identification of undigested remains was conducted from bird feathers, amphibian bones and skins and from chitin shells for insects. The identification of mammal species was based on both macroscopic examination of the teeth and microscopic examination of hairs following determination criteria (see Keller 1981a-c; Debrot et al. 1982; Erôme & Aulagnier 1982). The standardized niche breadth index of Levins (Levins 1968 *in* Krebs 1999) was used to compare the degree of food habit specialization per types of cat, by taking into account the food categories considered by Stahl (1986), Fitzgerald (1988) and Biró et al. (2005): 'small-size mammals', 'birds', 'other animals' (amphibians, insects) and 'human-linked food'. The index of Levins ranges from 0 (very specialized food habits) to 1 (very generalized food habits). The degree of similarities between the food habits of wildcats, domestic cats and hybrids was also estimated by using the Niche overlap index (Pianka 1973) ranging from 0 (no overlap) to 1 (full overlap).

We investigated the spatial-temporal sharing between sympatric wildcats, domestic cats and hybrids with radiotracking (Germain et al. 2008). The field study was carried out from May 2005 to December 2006 in a 1450 ha rural area located in North-Eastern France where wildcats have been present for centuries. The study area includes a small village closed to forest and surrounded by a mosaic of pastures, cultures and grasslands, and also by three farms. According to Pierpaoli et al. (2003), this area offers a very suitable habitat for wildcats, due to the forest continuity, as well as a potential landscape context where hybridization might occur because of the close proximity between forest and villages or farms.

Cats were trapped and fitted with 50g radio-collars with activity sensor. Hairs were collected for genotype determination (see O'Brien et al. 2009). Radio-tracking was conducted in attempt to obtain a minimum of 30 fixes per individual per season. Individual seasonal home-ranges were estimated with the 95% kernel with least-squares-cross validation (Worton 1989; Seaman & Powell 1996). Linear mixed models (fitted by maximum likelihood and with individuals as random factor; Pinheiro & Bates 2000) were used to investigate the relationship between seasonal home-ranges sizes and the type of cats, the sex, the body mass, the season, and the mating period. We also compared the proportion of active locations between the three types of cats. We computed an overlap index to estimate the degree of concordance/similarity between individual space use patterns. This index estimates the degree of similarity between two space use distributions (see Schmid & Schmidt 2006). It states that any location is characterized by a value equal to the lower of the two local activity densities occurring there, with these minimal values being assumed to reflect the likelihood of encounter between the two focal individuals. The overlap index ranges from 0 (no area shared) to 1 (identical same space use patterns).

Is the Behaviour of Hybrids Intermediate between Wildcat and Domestic Cat Behaviours?

Our results are in line with those of Moléon & Gil-Sanchez (2003), Malo et al. (2004), Lozano et al. (2006) and Piñeiro & Barja (2011) about the fact that wildcats have a 'facultative specialist' diet. In our studies, wildcats are 'specialists' in that their diet is mainly based on one type of prey (the small rodents) in accordance with results of other French studies from Condé et al. (1972) and Riols (1988). They are 'facultative specialists' in that they eat wide-ranging species of rodents. The presence of small wild rodents and small rodents commensal with humans, anthropic food, insects and birds (sparrows or chickens) in the domestic cat stomachs confirms the generalist character of the domestic cat's diet, which has already been demonstrated in other studies (e.g. Fitzgerald 1988; Barratt 1997b; Weber & Dailly 1998).

More precisely, and in accordance with Biró et al. (2005), our results suggest an important likeness between the food habits of wildcats, hybrids and domestic cats in regards to their frequent predation on small-sized mammals (rodents and insectivores). This likeness is more particularly evident between hybrids and domestic cats as they also eat amphibians, insects and anthropic food (wet and dry cat food, table scraps). The intermediate character of hybrids' diet, close to wildcats but relatively generalized like domestic cats, is described both by Biró et al. (2005) in Hungary and by us in France, hundreds of kilometres away. If generalizable to other hybrid populations, the behavioural flexibility of hybrids' diet could make them less sensitive than wildcats to fluctuations in the availability of preys.

In contrast with the results by Biró et al. (2005), we found that hybrids are at an intermediate degree of specialization between wildcats (narrowest trophic niche) and domestic cats (broadest trophic niche), but with food habits closer to those of domestic cats than to wildcat food habits. This observation would seem to indicate that hybrids and domestic cats may have more chances of meeting than wildcats and domestic cats have. As wildcats, hybrids seem to consume forest rodents more frequently than the meadow rodents. Thus, wildcats and hybrids may have more chances of meeting in forests than in meadows. The two types of cats may also approach human settlements, as indicated by the presence of house mice in their stomachs. In Hungary, Biró et al. (2005) found the House mouse (*Mus musculus)* only in the stomachs of hybrids and domestic cats while in France we found it in the stomachs of both types of cats. Nevertheless, considering the relative occurrence of human-linked food in the stomachs of hybrids and domestic cats, meetings in the vicinity of human settlements most probably involve hybrids and domestic cats. Further investigations are needed to see how similarities between food habits may really contribute to meetings - and thus potential interbreeding - between the three types of cats.

The mean size of the seasonal home-ranges of the five wildcats (four females and one male), the four hybrids (two females and two males) and the 13 domestic cats (seven females and six males) we radio-tracked range from 122 to 404 ha, from 76 to 358 ha and from 2 to 220 ha respectively. Mean home-range sizes of hybrids and wildcats did not differ in size whereas, like in Corbett (1979) and Biró et al. (2004), the mean home-range sizes of domestic cats were generally smaller than those of sympatric wildcats and related to the body mass. Biró et al. (2004) argued that wildcats have larger home-ranges than domestic cats because they are heavier, but the three types of cats had similar mean body masses in our study. As differences in mean body mass and size are assumed to affect the success of interbreeding between cats from different taxa (Leyhausen 1979), we supposed that these criteria were unlikely to act as hybridization barriers in our study area. Wildcats, as well as hybrids, may need larger home-ranges to compensate for the pluri-annual and seasonal variations in the abundance of small rodents (Stahl et al. 1988; Liberek 1999). In contrast, domestic cats easily find food and shelters in human settlements (Liberg & Sandell 1988) and thus, they do not require large home-ranges to ensure their safety and energy needs.

Contrary to the studies conducted by Corbett (1979) and Stahl et al. (1988) on European wildcats, we found that the home-range sizes of wildcats and hybrids did not vary with seasons and reproductive periods. In contrast, the home-ranges of domestic cats were smaller during colder periods (autumn to spring) than in summer. The relative confinement of domestic cats inside farm buildings during the cold months, when wildcats display reproductive activity, could limit opportunities for hybridization in our study area. Like Szemethy (1993) and Barratt (1997a), we observed that the home-ranges of domestic cats were centred on farms. Unsurprisingly, the higher overlap values were for these cats, confirming that farm cats have a communal living leading to a strong overlap between the home-ranges of individuals sharing the same farm (Liberg & Sandell 1988). The concordance in space use pattern between male and female wildcats was low but not null, at least in spring (i.e. mating season) and summer. Similarly, space use patterns of male and female hybrids overlapped at least to a small extent during summer.

Temporal segregation did not constitute a barrier to hybridization in our study area since the three types of cats have similar activity rhythms. Wildcats appeared to be mainly active at night, as already shown by Stahl et al. (1988), but were also active during the day, in accordance with the results of Scott et al. (1993). Wildcats, hybrids and domestic cats thus have opportunities to come into contact during nocturnal as well as during diurnal periods. However, we found almost no inter-sex overlap indexes between wildcats and domestic cats. Our results confirm the personal observations made by Klar et al. (2008) of a separation between home-ranges of domestic cats and wildcats in Germany, whereas Sarmento et al. (2009) observed that areas formerly used by wildcats have been replaced by domestic cats in Portugal.

Our results indicate that cats of opposite sexes have more chances of meeting a partner of their own type than one of another type. A high degree of concordance between the space use patterns of wildcats and domestic cats is therefore not a necessary prerequisite for hybridization. This suggests that even if hybridization occurred in our study area, meetings between domestic cats and wildcats of the opposite sex are probably relatively rare. They may occur through rare excursions of domestic cats far from their home-ranges. Such excursions are described by Say & Pontier (2004) and Yamane et al. (1996) in populations of urban or feral cats and can occur during dispersion. Two domestic cats we radio-tracked moved far away from the farm where they were captured (1500 m for a female and 2500 m for a male). The same male shifted from one farm to another during a winter night but came back to the first farm the next autumn.

Such exceptional excursions also occurred in strictly territorial mammals during breeding seasons (e.g. the American red squirrel, *Tamiasciurus hudsonicus*, see Benhamou 1996). Thus, wildcats may also make excursions far from their usual home-ranges during mating season. Males generally disperse farther and more rapidly than females (Hubbard et al. 1992). Females being more philopatric, male wildcats and male domestic cats may more often penetrate home-ranges of the other type of cats for mating with females than the other way around (Hertwig et al. 2009). In Hungary, home-ranges of male wildcats may cover farms and female domestic cat home-ranges (Szemethy 1993 during breeding season; Biró et al. 2004 during march-july period). In the French Ardennes Region, we also observed wildcats and hybrids getting near farms and villages: one female wildcat was observed four different times at night in pastures at least 100m from houses (twice in summer and twice in autumn), two hybrid males were also located in pastures at least 100m from a farm (twice in summer and once in the winter).

Finally, our results show that hybridization may occur even if wildcats and domestic cats do not necessarily exploit shared areas. They also provide information concerning the way the introgression of hybrid genes into wildcat populations may occur. A minimal degree of concordance in space use pattern between male hybrids and female wildcats, or between hybrids and domestic cats, of both sexes existed all year long (especially in summer), even if the overlap indexes are low. Our direct observations of interactions between cats of different types and opposite sex concerned a hybrid male and a wildcat female, which were once seen at less than 30m apart during the wildcat reproductive period, and a domestic male and a hybrid female, which were seen at less than 15m apart in summer. It seems that just because of their space use patterns, male and female hybrids have more opportunities for potential interbreeding with wildcats as well as with domestic cats, than wildcats and domestic cats have.

Possible Direction of Hybridization

Even though they are sparse, the few observations relative to the possible direction of hybridization made over the course of this study should be pointed out since little data is available on this subject. The four hybrids we radio-tracked had a 'wildcat phenotype' (see first paragraph) and, as the five wildcats we also radio-tracked, they were captured in traps located in forests, whereas the 13 fitted domestic cats were captured in traps located on farms or their surrounding areas. By contrast with domestic cats, the four hybrids were thus trapped in the same habitat as wildcats.

More generally, our data indicate that hybrids tend to behave like wildcats, be it for feeding habits or spatial behaviour. From these observations we could assume that the hybrids tracked do not come from female domestic cats and, under the hypothesis that they are first generation (F1 hybrids), they were raised and socialized by a female wildcat and not a female domestic cat. Because of this, they should not have any problems integrating into the social organization of wildcats, if we considered the example of hybrids resulting from mating between Ethiopian wolves (*Canis simensis*) and dogs (*C. familiaris*). Indeed, in Ethiopian wolves, during the reproduction period, females leave their pack to mate with neighbouring males (Sillero-Zubiri & Gotteli 1995). As there are few conspecific males, there is a high probability of mating with a male dog (Gottelli et al. 1994). Once mating has taken place, females raise their hybrid offspring with the natal pack members. Hybrids raised in a pack do not have, in theory, problems integrating other packs later on (Vilà & Wayne 1999). Thus, as European wildcats are solitary and as males do not assist the rearing or care of the offspring (Stahl & Léger 1992), the spatial behaviour of the hybrids we radio-tracked suggests that these hybrids were brought up and socialized by female wildcats.

The fact that these four hybrids had, theoretically, a wildcat mother also leads to the belief that hybridization takes place mainly in the direction of male domestic cat x female wildcat, which confirms the hypothesis of Oliveira et al. (2007) that states that the introgression of genes is unidirectional with only a flow of domestic genes into the wildcat populations. However, this observation goes against the statement made by Hubbard et al. (1992) that male wildcats probably mate with female domestic cats, and also against Eckert et al. (2010) who stated that the most probable scenario is that hybridization results from mating between male wildcats and female domestic cats. Our data tend more toward the idea that behavioural barriers are more effective for limiting mating between male wildcats and female domestic cats than for limiting mating between male domestic cats and female wildcats. This difference in effectiveness could be due to the fact that male wildcats have a period of sexual inactivity (see Condé & Schauenberg 1974; Corbett 1979) contrary to male domestic cats that are fertile throughout the year (see Artois et al. 2002 for a review). Thus, male wildcats may have fewer opportunities to mate with female domestic cats than male domestic cats have with female wildcats. Similarly, Vilà & Wayne (1999) think that dog-gray wolf mating occur mainly between male dogs and female gray wolves since the physiological aptitudes of male gray wolves are not synchronized with female dogs' oestrus. In the future, the development of studies on mitochondrial DNA of the various individuals (i.e. wildcats, domestic cats and hybrids) will help precisely identify the direction(s) of hybridization and introgression (Eckert et al. 2010; Say et al. 2011).

Hybridization: Threat or Chance for the Wildcat?

Establishing conservation measures for taxa first requires the definition of their taxonomic status (O'Brien & Mayr 1991; Brookes 1996). According to the biological concept of species (Mayr 1996), two species are different if they cannot reproduce together and produce fertile offspring. As soon as it is realized that two species are reproducing together, the initially protected species can lose its species status as well as its legal protection status (Brookes 1996; Arnold 1997). Hybridization between wildcats and domestic cats that produce fertile hybrids (Ragni 1993) could thus have important implications regarding to the conservation status of this wild felid (Balharry et al. 1997; Daniels et al. 1998) by directing the debate on the traditional vision of a 'reproductively isolated' species as a unit of conservation. Daniels & Corbett (2003) argue that "a new approach is necessary for defining mammalian species in the face of introgression with their domestic forms …" considering that the European wildcat still exists, but probably in a different form than its ancestors. As we have seen throughout this chapter, it is possible that a progressive shift in wildcat ecology and behaviour towards a 'hybrid ecology and behaviour' inherited from both domestic and wild ancestors could occur. However, it is not because the wildcat has no longer been reproductively isolated from the domestic cat for thousands of years and that its 'type' is difficult to determine, as it is probably still evolving, that populations of the wild feline should not be protected. This idea is in line with Daniels & Corbett (2003), who consider that "protecting animals per 'type' is illogical and impractical" whereas considering the functional role of European wildcats in ecosystems will be the way forward for maintaining of its conservation status. This idea is reinforced by the fact that as we have seen, there is no reliable and exclusive criterion that can help distinguish wildcats from hybrids in the field (Ruette et al. 2011).

The conservation of the functional roles would be all the more relevant since, once hybridization has begun, it is difficult to control, particularly if hybrids are fertile and reproduce together or with parental populations. In this case of introgressive hybridization, a veritable swarm of hybrids could form and in successive generations the number of hybrids would progressively increase (Allendorf et al. 2001). Reintroduction may be planned for maintaining the parental populations when 'widespread introgression' occurs ("Type 5" in Allendorf et al. 2001). Another way for conservation is the protection of the hybrids themselves, which may be the only solution when 'complete admixture' occurs ("Type 6" in Allendorf et al. 2001). The place and role of hybrids in wildcat populations should thus be considered as they look like wildcats and behave like wildcats. The concept of having two groups (wildcats and hybrids) is perhaps too academic, given their similarities and coexistence. This idea is reinforced by the fact that the potential hybridization of wildcat populations is less significant if it is considered that the wildcat belongs to a polytypic species than if it is considered as a distinct species from the domestic cat (Nowell & Jackson 1996). By pushing this way of thinking to the extreme, hybrids could have contributed, or still contribute, to the reinforcement and the preservation of wildcat populations by diversifying and enriching them through individuals that adapt better to environmental modifications.

Hybridization significantly influences the evolution, dispersion and distribution of numerous animal species (Grant & Grant 1992; Rhymer & Simberloff 1996; Dowling & Secor 1997) even if the importance of its involvement in the evolution of animal species remains a topic of debate (see notably Dowling & Secor 1997; Arnold 1992, 1997; Barton 2001). Opinions diverge in particular about the positive or negative aspects of its long-term effects on parental populations, whatever its natural or anthropogenic origins (Rhymer & Simberloff 1996; Wolf et al. 2001; Allendorf et al. 2001). It turns out that the consequences of hybridization on parental populations depend notably on the survival of hybrids and their ability to reproduce (Arnold 1992; Arnold & Hodges 1995; Marshall et al. 2002). By reproducing together or with one or both parental taxa, viable and fertile hybrids help exchange genes between existing lineages, and therefore contribute to the emergence of new recombinant taxa (Barton 2001; Grant & Grant 1992). As in plants (Rieseberg 1997), it appears to be more and more a reality in the animal kingdom, particularly in vertebrates (Dowling & DeMarais 1993; Schwenk et al. 2008).

Hybrids could present new gene combinations (Arnold 1992) and, because of that, be better adapted to intermediary habitat or climate changes (Anderson 1948; Lewontin & Birch 1966; Dowling & Secor 1997). Hybridization may thus be especially useful for supporting the dynamic of small threatened populations. For example, the Florida panther (*Puma concolor coryi*) population was down to only around 50 individuals. The population suffered from a high level of inbreeding and was close to extinction (Roelke et al. 1993) until individuals from a genetically close population located in Texas were introduced (Gross 2005). Today, it would seem that the survival of new-borns increases with the degree of admixture (Hostetler et al. 2010). The same operation was also carried out in 1978 by introducing hybrids from mating between Siberian tiger (*Panthera altaica*) and Indian tiger (*P. tigris*) to increase the population of Indian tigers (see Wayne & Brown 2001).

Our data tend towards showing that hybridization could be considered as a source of variations for adapting to new environments since the hybrids we studied have more flexible behaviour than wildcats, particularly concerning their feeding habits. Thus, not only is a new approach required to define mammal species confronted with introgression (Daniels & Corbett 2003), but the implication of hybrids in conservation programmes should also be discussed in future conservation plans. As stated by Boitani (1984), it should be recalled that one of the first priorities of conservation biology is "(...) to maintain the potentialities for evolutionary processes rather than to freeze the contemporary picture of the biological world". By taking into account the potential role of hybrids in wildcat populations conservation management, conservationists will recognize that "contemporary populations represent one end of a morphological, genetic and/or ecological cline" as Daniels & Corbett (2003) advocate.

Conservation Perspectives

McOrist and Kitchener (1994) believe that wildcats reintroduction programmes are not necessary unless if populations went extinct. By contrast, Randi (2003), Klar et al. (2008) and Eckert et al. (2010) highlight the interests of reintroduction projects to promote natural recolonization processes of wildcat populations. The idea is to use large populations of wildcats as sources of individuals that may be released in areas where habitats are currently favourable for their installation. Establishing reintroduction programmes for wildcats or reinforcing their populations could help support re-colonization mechanisms already underway, but also avoid the extinction of isolated populations (Randi 2003). Several projects have been carried out in Europe, notably in Switzerland (Hartmann 2005), Germany (in Bavaria, Böhle & Kleisinger 2005) and Spain (Ruiz-Olmo & Mino 1993; Such et al. 2005). The experiment in Germany was successful (Breitenmoser & Breitenmoser-Würsten 2002) but, overall, the impact of these reintroductions on population dynamics of wildcats, and on hybridization, remains difficult to measure.

Some authors also recommend establishing public information campaigns on the specificity of the European wildcat, why it should be conserved and the importance of having domestic cats spayed or neutered and vaccinated (e.g. Hubbard et al. 1992; Stahl & Artois 1994; McOrist & Kitchener 1994; Daniels & Corbett 2003). Say et al. (2011) suggest identifying the direction of gene flow between wildcats and domestic cats using mitochondrial DNA techniques to identify the sex responsible for introgression and, in this case, recommend neutering these individuals. However, sterilization or excluding domestic cats remains a controversial topic. Even if sterilization campaigns were proposed (e.g. Daniels et al. 2001; Kitchener et al. 2005; Lecis et al. 2006), they represent a risk given the possibility of incorrect cats identification (Hubbard et al. 1992; Balharry et al. 1997). Currently, some authors recommend capturing and sterilizing domestic cats in certain regions of interest according to genetic units of wildcats (Daniels et al. 2001; Oliveira et al. 2007; Say et al. 2011). However, as previously discussed, the wildcat is sometimes hard to identify since there is a morpho-anatomical gradient from the wildcat to the domestic cat. Furthermore, it may live in varied environments that can be close to farms and villages. In that context, a decrease in domestic cat or hybrid populations through elimination campaigns could be harmful to the European wildcat, even when restraining this management to specific areas.

Such investments in European wildcat conservation could be vain if, at the same time, the laws on wildcat protection are not reinforced in the field and if large habitat surfaces favourable to wildcats (i.e. forest and scrub-pastureland mosaic of habitats) are not preserved (Stahl & Léger 1992; Balharry et al. 1997; Randi et al. 2001; Lozano et al. 2003; Lecis et al. 2006; Lozano et al. 2007). Habitat protection is especially important to favour the preservation of wildlife corridors that may help wildcats to move between habitat patches without having to cross anthropized sectors where the risks of mortality and meeting domestic cats are increased (Stahl & Artois 1994; Randi 2003; Klar et al. 2008, 2009; Hertwig et al. 2009; Say et al. 2011). Finally, a combination of several conservation measures is necessary. This combination should concern populations or areas where cats present the main phenotypic characteristics of the European wildcat (see first paragraph) and behave so that, despite possible slight differences (e.g. more flexibility), the functional role of European wildcat in ecosystems is globally preserved. The right combination of all these recommendations will probably be, long-term, the one that could reduce the risk of wildcat populations isolation and thus, the potential risk of regression or disappearance of European wildcats.

References

Allendorf F.W., Leary R.F., Spruell P. & Wenburg J.K. 2001. The problems with hybrids: setting conservation guidelines. *Trends in Ecology and Evolution*, 16: 613-622.

Anderson E. 1948. Hybridization of the habitat. *Evolution*, 2: 1-9.

Arnold M.-L. 1992. Natural hybridization as an evolutionary process. *Annual Review of Ecology and Systematics*, 23: 237-261.

Arnold M.-L. & Hodges S.A. 1995. Are natural hybrids fit or unfit relative to their parents? *Trends in Ecology and Evolution,* 10: 67-71.

Arnold M.-L. 1997. *Natural Hybridization and Evolution*. Oxford University Press. 215 p.

Artois M. 1985. Utilisation de l'espace et du temps chez le renard (*Vulpes vulpes*) et le chat forestier (*Felis silvestris*) en Lorraine. *Gibier Faune Sauvage*, 3: 33-57.

Artois M., Duchêne M.J., Pericard J.M. & Xemar V. 2002. Le chat domestique errant ou haret (*Felis silvestris catus*, Linnaeus, 1758). In: *Encyclopédie des Carnivores de France*. Société Française pour l'Etude et la Protection des Mammifères, Paris. 50 p.

Aymerich M. 1982. Etude comparative des régimes alimentaires du lynx pardelle (*Lynx pardina* Temminck, 1824) et du chat sauvage (*Felis silvestris* Schreber, 1777) au centre de la péninsule Ibérique. *Mammalia*, 46: 515-521.

Balharry E., Staines B.W., Marquiss M. & Kruuk H. 1994. *Hybridisation in British Mammals. JNCC Report No. 154.* Joint Nature Conservation Committee, Peterborough. 44p.

Balharry D., Daniels D. & Barratt E.M. 1997. Wildcats: can genetics help their conservation? In: *The role of genetics in conserving small population.* eds Tew T.E., Crawford T.J., Spencer J.W., Stevens D.P., Usher M.B. & Warren J. Peterborough, JNCC. pp. 102-111.

Balharry D. & Daniels M.J. 1998. Wild living cats in Scotland. Scottish Natural Heritage Research, *Survey and Monitoring Report*, n° 23. 83 p.

Barratt D.G. 1997a. Home range size, habitat utilization and movements patterns of suburban and farm cats *Felis catus*. *Ecography*, 20: 271-280.

Barratt D. G. 1997b. Predation by house cats, *Felis catus* (L.), in Canberra, Australia. I. Prey composition and preference. Wildlife Research 24, 263-277.

Barton N.H. 2001. The role of hybridization in evolution. *Molecular Ecology*, 10: 551-568.

Beaumont M., Barrat E.M., Gottelli D., Kitchener A.C., Daniels M.J., Pritchard J.K. & Bruford M.W. 2001. Genetic diversity and introgression in the Scottish wildcat. *Molecular Ecology*, 10: 319-336.

Benhamou S. 1996. Space use and foraging movements in the American red squirrel (*Tamiasciurus hudsonicus*). *Behavioural Processes*, 37: 89-102.

Biró Z., Szemethy L. & Heltai M. 2004. Home range sizes of wildcats (*Felis silvestris*) and feral domestic cats (*Felis silvestris* f. *catus*) in a hilly region of Hungary. *Mammalian Biology,* 69: 302-310.

Biró Z., Lanszki J., Szemethy L., Helti M. & Randi E. 2005. Feeding habits of feral domestic cats (*Felis catus*), wild cats (*Felis silvestris*) and their hybrids: trophic niche overlap among cat groups in Hungary. *Journal of Zoology*, 266: 187-196.

Böhle U.R. & Kleisinger H. 2005. 20 years of wildcat reintroduction in Bavaria: a validation by molecular techniques. In: *Oral presentation in symposium of biology and conservation of the European wildcat (Felis silvestris silvestris).* Germany, January 21st-23rd 2005.

Boitani L. 1984. Genetic considerations on wolf conservation in Italy. *Bolletino di Zoologia*, 51: 367-373.

Breitenmoser U. & Breitenmoser-Würsten C. 2002. Un bilan des réintroductions des carnivores en Europe. In: *L'étude et la conservation des carnivores.* Société Française pour l'Etude et la Protection des Mammifères, Paris. pp. 17-19.

Brookes M. 1996. *Hybridisation rules OK!* New Scientist, Forum, p. 56.

Bullini L. 1994. Origin and evolution of animal hybrid species. *Trends in Ecology and Evolution*, 9: 422-426.

Cabria M.T., Michaux J.R., Gómez-Moliner B.J., Skumatov D., Maran T., Fournier P., Lopez de Luzuriaga J. & Zardoya R. 2011. Bayesian analysis of hybridization and introgression between the endangered European mink (*Mustela lutreola*) and the polecat (*Mustela putorius*). *Molecular Ecology*, 20: 1176-1190.

Carr S.M., Ballinger S.W., Derr J.N., Blankenship L.H. & Bickham J.W. 1986. Mitochondrial DNA analysis of hybridization between sympatric while-tailed deer and mule deer in west Texas. *Proceedings of the National Academy of Sciences of the United States of America*, 83: 9576-9580.

Condé B. & Schauenberg P. 1969. Reproduction du chat forestier d'Europe (*Felis silvestris* Schreber) en captivité. *Revue Suisse de Zoologie*, 76: 183-210.

Condé B. & Schauenberg P. 1971. Le poids du chat forestier d'Europe (*Felis silvestris* Schreber 1777). *Revue Suisse de Zoologie*, 78: 295-315.

Condé B. & Schauenberg P. 1974. Reproduction du Chat forestier (*F. silvestris* Schr.) dans le nord-est de la France. *Revue Suisse de Zoologie*, 81: 45-52.

Condé B., Nguyen-Thi-Thu-Cuc, Vaillant F. & Schauenberg P. 1972. Le régime alimentaire du Chat forestier (*F. silvestris* Schr.) en France. *Mammalia*, 36: 112-119.

Corbett L.K. 1978. Current research on wildcats: why have they increased? *Scottish Wildlife*, 14: 17-21.

Corbett L.K. 1979. *Feeding ecology and social organization of wildcats (Felis silvestris) and domestic cats (Felis catus) in Scotland.* PhD Thesis, University of Aberdeen. 295 p.

Council of Europe. 1979. Convention on the conservation of European wildlife and natural habitats. Bern, 19.IX.1979. European Treaty Series n° 104.

Council of Europe. 1992. Directive 92/43/EEC of 21st May 1992 on the conservation of natural habitats and of wild fauna and flora. Appendix IV. Amended by: Council directive 2006/105/EC of 20th November 2006. European Commission Environment, accessed January 8th 2012: *http://ec.europa.eu/environment/nature/legislation/habitatsdirective/index_en.htm.*

Daniels M.J., Balharry D., Hirst D., Kitchener A.C. & Aspinall R.J. 1998. Morphological and pelage characteristics of wild living cats in Scotland: implications for defining the "wildcat". *Journal of Zoology*, 44: 231-247.

Daniels M.J., Beaumont M.A., Johnson P.J., Balharry D., Macdonald D.W. & Barratt E. 2001. Ecology and genetics of wild-living cats in the north-east of Scotland and the implications for the conservation of the wildcat. *Journal of Applied Ecology*, 38: 146-161.

Daniels M.J. & Corbett L. K. 2003. Redefining introgressed protected mammals: when is a wildcat a wild cat and a dingo a wild dog? *Wildlife research*, 30: 213-218.

Davison A., Birks J.D.S., Griffiths H.I., Kitchener A.C., Biggins D. & Butlin R.K. 1999. Hybridization and phylogenetic relationship between polecats and domestic ferrets in Britain. *Biological Conservation*, 87: 155-161.

de Planhol X. 2004. *Le paysage animal. L'homme et la grande faune : une zoogéographie historique*. Librairie Arthëme Fayard, Poitiers. 1127 p.

Debrot, S., Fivaz, G., Mermod, C., Weber, J.-M. 1982. Atlas des poils de mammifères d'Europe. *Institut Zoologique de l'Université de Neuchâtel, Neuchâtel.*

Dötterer M. & Bernhart F. 1996. The occurrence of wildcats in the southern Swiss Jura Mountains. *Acta Theriologica*, 41: 205-209.

Dowling T.E. & DeMarais B.D. 1993. Evolutionary significance of introgressive hybridization in cyprinid fishes. *Nature*, 362: 444-446.

Dowling T.E. & Secor C.L. 1997. The role of hybridization and introgression in the diversification of animals. *Annual Review of Ecology and Systematics,* 28: 593-619.

Driscoll C.A., Menotti-Raymond M., Roca A., Hupe K., Johnson W.E., Gelfen E., Harley E.H., Delibes M., Pontier D., Kitchener A.C., Yamaguchi N., O'Brien S.J. & Macdonald D.W. 2007. The near eastern origin of cat domestication. *Science*, 317: 519-523.

Driscoll C. & Nowell K. 2010. *Felis silvestris*. In: IUCN 2011. IUCN Red List of Threatened Species. Version 2011.2. <*www.iucnredlist.org*>. Downloaded on 08 January 2012.

Dudley J.P., Ginsberg J.R., Plumptre A.J., Hart J.A. & Campos L.C. 2002. Effects of war and civil strife on wildlife and wildlife habitat. *Conservation Biology,* 16: 319-329.

Easterbee N. 1991. Wildcat. In: *The Handbook of British Mammals*. 3rd edn. eds G. B. Corbet & S. Harris. Blackwell Scientific Publications, Oxford, pp. 431-37.

Eckert I. & Hartl G.B. 2005. Conservation genetics of the European wildcat (*Felis silvestris silvestris*) in Germany. In: *Poster presentation in symposium of biology and conservation of the European wildcat (Felis silvestris silvestris).* January 21st-23rd 2005, Germany.

Eckert I., Suchentrunk F., Markov G. & Hartl G.B. 2010. Genetic diversity and integrity of German wildcat (*Felis silvestris*) populations as revealed by microsatellites, allozymes, and mitochondrial DNA sequences. *Mammalian Biology*, 75: 160-174.

Erôme, G., Aulagnier, S., 1982. Contribution à l'identification des proies des rapaces. *Bièvre.* 4, 129-135.

Fitzgerald B.M. 1988. Diet of domestic cats and their impact on prey populations. In: *The domestic cat: the biology of its behaviour*. eds Turner D.C. & Bateson P. Cambridge University Press, Cambridge. pp. 123-145.

French D.D., Corbett L.K. & Easterbee N. 1988. Morphological discriminant of Scottish wildcats (*Felis silvestris*), domestic cats (*Felis catus*) and their hybrids. *Journal of Zoology*, 214: 235-259.

Gaston K.J. 2003. *The structure and dynamics of geographic ranges*. Oxford University Press, Oxford. 266 p.

Germain E., Benhamou S. & Poulle M.-L. 2008. Spatio-temporal sharing between the European wildcat (*Felis silvestris*), the domestic cat (*Felis catus*), and their hybrids. *Journal of Zoology*, 276: 195-203.

Germain E., Ruette S. & Poulle M.-L. 2009. Likeness between the food habits of European wildcats, domestic cats and their hybrids in France. *Mammalian Biology*, 74: 412-417.

Gottelli D., Sillero-Zubiri C., Applebaum G.D., Roy M.S., Girman D.J., Garcia-Moreno J., Ostranders E.A. & Wayne R.K. 1994. Molecular genetics of the most endangered canid: the Ethiopian wolf *Canis simensis*. *Molecular Ecology*, 3: 301-312.

Grant P.R. & Grant B.R. 1992. Hybridization in bird species. *Science*, 256: 193-197.

Gross L. 2005. Why not the best? How science failed the Florida panther? *PLos Biology,* 3: 1-7.

Hailer F. & Leonard J.A. 2008. Hybridization among three native North American *Canis* species in a region of natural sympatry. *PloS ONE*, 3: 1-9.

Haltenorth T. 1957. Die Wildkatze. *Die Neue Brehm-Bücherei* 189. A. Ziemsen Verlag. Wittenberg Lutherstadt. 101 p.

Hansen, C. M., 2010. *Movements and Predation Activity of Feral and Domestic Cats (Felis catus) on Banks Peninsula*. Lincoln University, Christchurch, New Zealand, 122p.

Hartmann M. 2005. Reproduction and behaviour of European wildcats in species-specific enclosures. In: *Oral presentation in symposium of biology and conservation of the European wildcat (Felis silvestris silvestris)*. January 21st-23rd 2005, Germany.

Heltai M., Biró Z. & Szemethy L. 2006. The changes of distribution and population density of wildcats *Felis silvestris* Schreber, 1775 in Hungary between 1987-2001. *Nature Conservation*, 62: 37-42.

Hertwig S.T., Schweizer M., Stepanow S., Jungnickel A., Böhle U.R. & Fischer. 2009. Regionally high rates of hybridization and introgression in German wildcat populations (*Felis silvestris*, Carnivora, Felidae). *Journal of Zoology Systematics and Evolutionnary Research*. 47: 283-297.

Hubbard A.L., McOrist S., Jones T.W., Boid R., Scott R. & Easterbee N. 1992. Is survival of European wildcats *Felis silvestris* in Britain threatened by interbreeding with domestic cats? *Biological conservation*, 61: 203-208.

Hostetler J.A., Onorato D.P., Nichols J.D., Johnson W.E., Roelke M.E., O'Brien S.J., Jansen D. & Oli M.K. 2010. Genetic introgression and the survival of Florida panther kittens. *Biological Conservation*, 143: 2789-2796.

Jackson P., Jackson A.F., Dallet R. & de Crem J. 1996. *Les félins - toutes les espèces du monde*. Delachaux & Niestlé, Lausanne. 272 p.

Jaeger J.A.G., Bowman J., Brennan J., Fahring L., Bert D., Bouchard J., Charbonneau N., Frank K., Gruber B. & Tluk von Toschanowitz K. 2005. Predicting when animal populations are at risk from roads: an interactive model of road avoidance behavior. *Ecological Modelling*, 185: 329-348.

Jerosch S., Götz M., Klar N. & Roth M. 2010. Characteristics of diurnal resting sites of the endangered European wildcat (*Felis silvestris silvestris*): Implications for its conservation. *Journal for Nature Conservation*, 18: 45-54.

Journal Officiel de la République Française. 1976. Loi n°76-629 du 10 juillet 1976 relative à la protection de la nature. *Journal Officiel de la République Française*, 13 juillet 1976: 4203-4206.

Keller A. 1981a. Détermination des mammifères de la Suisse par leur pelage: II. Diagnose des familles. III. Lagomorpha, Rodentia. *Revue Suisse de Zoologie,* 87: 781-796.

Keller A. 1981b. Détermination des mammifères de la Suisse par leur pelage: IV. Cricetidae et Muridae. *Revue Suisse de Zoologie*, 88: 463-473.

Keller A. 1981c. Détermination des mammifères de la Suisse par leur pelage: V. Carnivora, VI Artiodactyla. *Revue Suisse de Zoologie*, 88: 803-820.

Kelly E.G. & Forsman E.D. 2004. Recent records of hybridization between barred owls (*Strix varia*) and northern spotted owls (*S. occidentalis caurina*). *The Auk,* 121: 806-810.

Kitchener A.C. 1991. *The natural history of the wild cats*. Cornell University Press, New York. 280 p.

Kitchener A.C., Yamaguchi N., Ward J.M. & Macdonald D.W. 2005. A diagnosis for the Scottish wildcat (*Felis silvestris*): a tool for conservation action for a critically-endangered felid. *Animal Conservation*, 8: 223-237.

Kitchener A.C. & Rees E.E. 2009. Modelling the dynamic biogeography of the wildcat: implications for taxonomy and conservation. *Journal of Zoology,* 297: 144-155.

Klar N. Fernández N., Kramer-Schadt S. Hermann M., Trinzen M., Büttner I. & Niemitz C. 2008. Habitat selection models for European wildcat conservation. *Biological Conservation*, 141: 308-319.

Klar N., Herrmann M. & Kramer-Schadt S. 2009. Effects and mitigation of road impacts on individual movement behavior of wildcats. *The Journal of Wildlife Management*, 73: 631-638.

Krüger M., Hertwig S.T., Jetschke G. & Fischer M.S. 2009. Evaluation of anatomical characters and the question of hybridization with domestic cats in the wildcat population of Thuringia, Germany. *Journal of Zoology Systematics and Evolutionnary Research*, 47. 268-282.

Kohn M, Knauer F, Stoffella A, Schröder W & Pääbo S. 1995. Conservation genetics of the European brown bear: a study using excremental PCR of nuclear and mitochondrial sequences. *Molecular Ecology*, 4: 95-103.

Krebs C.J. 1999. *Ecological methodology*. 2nd Edition. Addison-Wesley, New-York. 620p.

Lecis R., Pierpaoli M., Biró Z.S., Szemethy L., Ragni B., Vercillo F. & Randi E. 2006. Bayesian analyses of admixture in wild and domestic cats (*Felis silvestris*) using linked microsatellite loci. *Molecular Ecology*, 15: 119-131.

Léger F., Stahl P., Ruette S. & Wilhelm J.L. 2008. La répartition du chat forestier en France : évolutions récentes. *Faune Sauvage*, 280: 24-39.

Lehman N., Eisenhawer A., Hansen K., Mech L.D., Peterson R.O., Gogan P.J.P. & Wayne R.K. 1991. Introgression of coyote mitochondrial DNA into sympatric north American Gray wolf populations. *Evolution*, 45: 104-119.

Levins R.1968. *Evolution in changing environments: Some theoretical explorations.* Princeton University Press, Princeton, New Jersey. 132p.

Lewontin R.C. & Birch L.C. 1966. Hybridization as a source of variation for adaptation to new environment. *Evolution*, 20: 315-336.

Leyhausen P. 1979. *Cat behavior. The predatory and social behavior of domestic and wild cats.* Garland Publishing, Inc. New York. 340 p.

Liberek M. 1999. Eco-éthologie du Chat sauvage *Felis s. silvestris*, Schreber 1777 dans le Jura vaudois (Suisse). *Influence de la couverture neigeuse*. PhD Thesis, Université de Neuchatel. 258 p.

Liberg O. & Sandell M. 1988. Spatial organisation and reproductive tactics in the domestic cat and other felids. In: *The domestic cat: the biology of its behaviour*. eds Turner D.C. & Bateson P. Cambridge University Press, Cambridge. pp. 83-98.

Lozano J., Virgós E., Malo A.F., Huertas D.L. & Casanovas J.G. 2003. Importance of scrub-pastureland mosaic for wild-living cats occurrence in a Mediterranean area: implications for the conservation of the wildcat (*Felis silvestris*). *Biodiversity and Conservation*, 12: 921-935.

Lozano J., Moleón M. & Virgós E. 2006. Biogeographical patterns in the diet of the wildcat, *Felis silvestris* Schreber, in Eurasia: factors affecting the trophic diversity. *Journal of Biogeography*, 33: 1076-1085.

Lozano J., Virgós E., Cabezas-Díaz S. & Mangas J.G. 2007. Increase of large game species in Mediterranean areas: Is the European wildcat (*Felis silvestris*) facing a new threat? *Biological Conservation*, 138: 321-329.

Mallet J. 2005. Hybridization as an invasion of the genome. *Trends in Ecology and Evolution*, 20: 229-237.

Malo A.F., Lozano J., Huertas D.L. & Virgós E. 2004. A change in the diet from rodents to rabbits (*Oryctolagus cuniculus*). Is the wildcat (*Felis silvestris*) a specialist predator? *Journal of Zoology*, 263: 401-407.

Maran T. & Henttonen H. 1995. Why is the European mink (*Mustela lutreola*) disappearing? A review of the process and hypotheses. *Annual Journal of Zoology*, 32: 47-54.

Marshall J.L., Arnold M.L. & Howard D.J. 2002. Reinforcement: the road not taken. *Trends in Ecology and Evolution,* 17: 558-563.

Mayr E. 1996. What is a species, and what is not? *Philosophy of Science,* 63: 262-277.

McOrist S. & Kitchener A.C. 1994. Current threats to the European wildcats, *Felis silvestris*, in Scotland. *Ambio*, 23: 243-245.

Mermod C.P. & Liberek M. 2002. The role of snowcover for European wildcat in Switzerland. *Zeitschrift für Jagdwissenschaft*, 48: 17-24.

Moleón M. & Gil-Sánchez J.M. 2003. Food habits of the wildcat (*Felis silvestris*) in a peculiar habitat: the Mediterranean high mountain. *Journal of Zoology*, 260: 17-22.

Monterroso P., Brito J. C., Ferreras P. & Alves P. C. 2009. Spatial ecology of the European wildcat in a Mediterranean ecosystem: dealing with small radio-tracking datasets in species conservation. *Journal of Zoology*, 279: 27-35.

Moore G.C. & Parker, G.R. 1992. Colonization by the Eastern coyote (*Canis latrans*). In: *Ecology and Management of the Eastern Coyote.* ed Boer A.H. *Wildlife Research Unit*, New Brunswick. pp. 23-38.

Nowell K. & Jackson P. 1996. *Wild cats: Status survey and conservation action plan.* eds Nowell K. & Jackson P. IUCN, Gland. 383 p.

O'Brien S.J. & Mayr E. 1991. Bureaucratic mischief: recognizing endangered species and subspecies. *Science*, 251: 1187-1188.

O'Brien S.J., Johnson W., Driscoll C., Pontius J., Pecon-Slattery J. & Menotti-Raymond M. 2008. State of cat genomics. *Trends in Genetics*, 24: 268-279.

O'Brien J., Devillard S., Say L., Vanthomme H., Léger F., Ruette S. & Pontier D. 2009. Preserving genetic integrity in a hybridising world: are European wildcats (*Felis silvestris silvestris*) in eastern France distinct from sympatric feral domestic cats? *Biodiversity and Conservation*, 18: 2351–2360.

Oliveira R., Godinho R., Randi E., Ferrand N. & Alves P.C. 2007. Molecular analysis of hybridisation between wild and domestic cats (*Felis silvestris*) in Portugal: implications for conservation. *Conservation Genetics,* 9: 1-11.

Oliveira R, Godinho R, Randi E, Alves PC. 2008. Hybridization versus conservation: are domestic cats threatening the genetic integrity of wildcats (*Felis silvestris silvestris*) in Iberian Peninsula? *Philosophical Transactions of the Royal Society B*, 363: 2953-61.

Orr M.R. & Smith T.B. 1998. Ecology and speciation. *Trends in Ecology and Evolution,* 13: 502-506.

Parent G.H. 1975. La migration récente à caractère invasionnel du chat sauvage, *Felis silvestris silvestris* Schreber, en Lorraine Belge. *Mammalia*, 39: 251-288.

Paterson B.D. 1999. Contingency and determinism in mammalian biogeography: the role of history. *Journal of Mammalogy*, 80: 345-360.

Peterson A.T. & Robins C.R. 2003. Using ecological-niche modelling to predict barred owl invasions with implications for spotted owl conservation. *Conservation Biology*, 17: 1161-1165.

Pianka E.R. 1973. The structure of lizard communities. *Annual Review of Ecology and Systematics,* 4: 53-74.

Pierpaoli M., Biró Z.S., Herrmann M., Hupe K., Fernandes M., Ragni B., Szemethy L. & Randi E. 2003. Genetic distinction of wildcat (*Felis silvestris*) populations in Europe, and hybridization with domestic cats in Hungary. *Molecular Ecology*, 12: 2585-2598.

Piñeiro A. & Barja I. 2011. Trophic strategy of the wildcat *Felis silvestris* in relation to seasonal variation in the availability and vulnerability to capture of Apodemus mice. *Mammalian Biology*, 76: 302-307.

Pinheiro J.C. & Bates D.M. 2000. *Mixed-Effects Models in S and S-PLUS.* Springer. 528 p.

Platz S., Hertwig S.T., Jetschke G., Krüger M. & Fischer M.S. 2011. Comparative morphometric study of the Slovakian wildcat population (*Felis silvestris silvestris*): Evidence for a low rate of introgression? *Mammalian Biology*, 76: 222-233.

Ragni B. 1993. Status and conservation of the wildcat in Italy. In: *Council of Europe 1992: Seminar on the biology and conservation of the wildcat (Felis silvestris).* Council of Europe, Strasbourg. pp. 40-41.

Ragni B. & Randi E. 1986. Multivariate analysis of craniometric characters in European wild cat, Domestic cat and African wild cat (genus *Felis*). *Zeitschrift fur Säugetierkunde,* 51: 243-251.

Ragni B. & Possenti M. 1996. Variability of coat-colour and marking system in *Felis silvestris*. *Italian Journal of Zoology,* 63: 285-292.

Randi E. 2003. Conservation genetics of carnivores in Italy. *Compte Rendus de Biologie*, 326: S54-S60.

Randi E. 2008. Detecting hybridization between wild species and their domesticated relatives. *Molecular Ecology*, 17: 285-93.

Randi E. & Ragni B. 1991. Genetic variability and biochemical systematics of domestic and wild cat populations (*Felis silvestris*: Felidae). *Journal of Mammalogy*, 72: 79-88.

Randi E., Pierpaoli M., Beaumont M., Ragni B. & Sforzi A. 2001. Genetic identification of wild and domestic cats (*Felis silvestris*) and their hybrids using bayesian clustering methods. *Molecular Biology and Evolution*, 18: 1679-1693.

Rhymer J.M. & Simberloff D. 1996. Extinction by hybridization and introgression. *Annual Review of Ecology and Systematics*, 27: 83-109.

Rieseberg L.H. 1997. Hybrid origins of plant species. *Annual Review of Ecology and Systematics*, **28:** 359-389.

Riols, C., 1988. Etude du régime alimentaire du Chat forestier (*Felis silvestris*) dans l'Est de la France. *Bulletin Mensuel de l'Office National de la Chasse*, 121: 22-27.

Roelke M.E., Martenson J.S. & O'Brien S.J. 1993. The consequences of demographic reduction and genetic depletion in the endangered Florida panther. *Current Biology*, 3: 340-350.

Ruette S., Germain E., Léger F., Say L. & Devillard S. 2011. Identification du chat forestier en France. Apport de la génétique pour détecter les "hybrides". *Faune Sauvage*, 292: 10-16.

Ruiz-Olmo J. & Mino A. 1993. Reintroduction of wildcats in Catalonia (NE Spain). In: *Council of Europe 1992: Seminar on the biology and conservation of the wildcat (Felis silvestris)*. Council of Europe, Strasbourg. pp. 111-114.

Sarmento P. 1996. Feeding ecology of the European wildcat *Felis silvestris* in Portugal. *Acta Theriologica*, 41: 409-414.

Sarmento P., Cruz J., Tarroso P. & Fonseca C. 2006. Space and habitat selection by female European wild cats (*Felis silvestris silvestris*). *Wildlife Biology in Practice*, 2: 79-89.

Sarmento P., Cruz J., Eira C. & Fonseca C. 2009. Spatial colonization by feral domestic cats *Felis catus* of former wildcat *Felis silvestris silvestris* home ranges. *Acta theriologica*, 54: 31-38.

Say L., Devillard S., Natoli E. & Pontier D. 2002. The mating system of feral cats (*Felis catus* L.) in a sub-Antarctic environment. *Polar Biology*, 25: 838-842.

Say L. & Pontier D. 2004. Spacing pattern in a group of stray cats: effects on male reproductive success. *Animal Behaviour*, 68: 175-180.

Say L., Devillard S., Léger F., Pontier D. & Ruette S. 2011. Distribution and spatial genetic structure of European wildcat in France. *Animal Conservation,* [Early View].

Schauenberg P. 1969. L'identification du Chat forestier d'Europe *Felis s. silvestris* Schreber 1777 par mesure ostéométrique. *Revue Suisse de Zoologie,* 76: 433-441.

Schauenberg P. 1977a. La stature du Chat forestier *Felis silvestris* Schreber, et la variabilité morphologique de l'espèce. *Revue Suisse de Zoologie*, 84: 323-337.

Schauenberg P. 1977b. Longueur de l'intestin du Chat forestier *Felis silvestris* Schreber. *Mammalia*, 41: 357-360.

Schmid F. & Schmidt A. 2006. Nonparametric estimation of the coefficient of overlapping - theory and empirical application. *Computational Statistics and Data Analysis*, 50: 1583-1596.

Schwartz M.K., Pilgrim K.L., McKelvey K.S., Lindquist E.L., Claar J.J., Loch S. & Ruggiero L.F. 2004. Hybridization between Canada lynx and bobcats: genetic results and management implications. *Conservation Genetics*, 5: 349-355.

Schwenk K., Brede N. & Streit B. 2008. Introduction. Extent, processes and evolutionary impact of interspecific hybridization in animals. *Philosophical Transactions of the Royal Society*, 363: 2805-2811.

Scott R., Easterbee N. & Jefferies D. 1993. A radio-tracking study of wildcats in western Scotland. In: *Council of Europe 1992: Seminar on the biology and conservation of the wildcat (Felis silvestris)*. Council of Europe, Strasbourg. pp. 94-97.

Seaman D.E. & Powell R.A. 1996. An evaluation of the accuracy of kernel density estimators for home range analysis. *Ecology*, 77: 2075-2085.

Seehausen O. 2004. Hybridization and adaptative radiation. *Trends in Ecology and Evolution,* 19: 198-207.

Serpell J.A. 1988. The domestication and the history of the cat. In: *The domestic cat: the biology of its behaviour*. eds Turner D.C. & Bateson P. Cambridge University Press, Cambridge. pp. 151-158.

Sillero-Zubiri C. & Gottelli D. 1995. Spatial organization in the Ethiopian wolf *Canis simensis*: large packs and small stable home ranges. *Journal of the Zoology*, 237: 65-81.

Silva-Rodriguez E. A. & Sieving K. E. 2011. Influence of care of domestic carnivores on their predation on vertebrates. *Conservation Biology*, 25: 808-815.

Sommer R.S. & Benecke N. 2006. Late Pleistocene and Holocene development of the Felid fauna (Felidae) of Europe: a review. *Journal of Zoology*, 269: 7–19.

Stahl P. 1986. Le Chat forestier d'Europe (*Felis silvestris*, Schreber 1777). *Exploitation des ressources et organisation spatiale*. PhD Thesis, Université de Nancy I. 356 p.

Stahl P., Léger F. & Artois M. 1984. *Quelques observations sur la reproduction du chat forestier (Felis silvestris) en Lorraine. Ciconia,* 8: 69-74.

Stahl P., Artois M. & Aubert M.F.A. 1988. Organisation spatiale et déplacements des chats forestiers adultes (*Felis silvestris*, Schreber, 1777) en Lorraine. *Revue d'Ecologie* (Terre &Vie), 43 : 113-132.

Stahl P. & Léger F. 1992. Le chat sauvage (*Felis silvestris*, Schreber, 1777). In: *Encyclopédie des Carnivores de France. Société Française pour l'Etude et la Protection des Mammifères*, Paris. 50 p.

Stahl P. & Artois M. 1994. *Status and conservation of the wildcat (Felis silvestris) in Europe and around the Mediterranean rim. Nature and Environment Series*, N° 69. Council of Europe Press, Strasbourg. 78 p.

Such A., López-Martín J.M., Martínez D. & Pinyol C. 2005. Effect of different captive environments and release techniques of dispersion movements of releases captive-reared wildcats (*Felis silvestris*). In: *Oral presentation in symposium of biology and conservation of the European wildcat (Felis silvestris silvestris)*. January 21st-23rd 2005, Germany.

Suminski P. 1962. Les caractères de la forme pure du chat sauvage *Felis silvestris* Schreber. *Archives des Sciences*, 15 : 277-297.

Sunquist M. & Sunquist F. 2002. *Wild cats of the world*. The University of Chicago Press, Chicago. 452 p.

Szemethy L. 1993. The actual status of the wildcat (*Felis silvestris*). In: *Council of Europe 1992: Seminar on the biology and conservation of the wildcat (Felis silvestris)*. Council of Europe, Strasbourg. pp 52.

Turner, D.C. & Bateson, P. 1988. *The domestic cat: the biology of its behaviour*. Cambridge University Press, Cambridge. 222 p.

Tryjanowski P., Antczak M., Hromada M., Kuczynski L. & Skoracki M. 2002. Winter feeding ecology of male and female European wildcats *Felis silvestris* in Slovakia. *Zeitschrift für Jagdwissenschaft*, 48: 49-54.

Tschanz B., Hegglin D., Gloor S., Bontadina F. 2011. Hunters and non-hunters: skewed predation rate by domestic cats in a rural village. *European Journal of Wildlife Research*, 57: 597-602.

UNEP-WCMC. 8 January, 2012. *UNEP-WCMC Species Database: CITES-Listed Species* Available at *http://www.cites.org/fra/resources /species.html.*

Verardi A., Lucchini V. & Randi E. 2006. Detecting introgressive hybridization between free-ranging domestic dogs and wild wolves (*Canis lupus*) by admixture linkage disequilibrium analysis. *Molecular Ecology*, 15: 2845-2855.

Vigne J.-D., Guilaine J., Debue K., Haye L. & Gérard P. 2004. Early taming of the cat in Cyprus. *Science*, 304: 259.

Vigne J.-D. & Guilaine J. 2004. Les premiers animaux de compagnie, 8500 ans avant notre ère ? ...ou comment j'ai mangé mon chat, mon chien et mon renard. *Anthropozoologica*, 39: 249-273.

Vilà C. & Wayne R.K. 1999. Hybridization between wolves and dogs. *Conservation Biology*, 13: 195-198.

Wayne R.K. & Brown D.M. 2001. Hybridization and conservation of carnivores. In: *Conservation Biology 5 - Carnivore Conservation*. eds Gittleman J.L., Funk S.M., Macdonald D. & Wayne R.K. Cambridge University Press, Cambridge. pp. 145-162.

Weber, J.-M. & Dailly L. 1998. Food habits and ranging behaviour of a group of farm cats (*Felis catus*) in a Swiss mountainous area. *Journal of Zoology*, 245: 234-237.

Wittmer H.U. 2001. Home range size, movements, and habitat utilization of three male European wildcats (*Felis silvestris* Schreber, 1777) in Saarland and Rheinland-Pfalz (Germany*). Mammalian Biology*, 66: 365-370.

Wolf D.E., Takebayashi N. & Rieseberg L.H. 2001. Predicting the risk of extinction through hybridization. *Conservation Biology,* 15: 1039-1053.

Wooding S. & Ward R. 1997. Phylogeography and Pleistocene evolution in the North American black bear. *Molecular Biology and Evolution*, 14: 1096-1105.

Worton B.J. 1989. Kernel methods for estimating the utilization distribution in home-range studies. *Ecology*, 70: 164-168.

Yamaguchi N., Kitchener A.C., Driscoll C.A., Ward J.M. & Macdonald D.W. 2004. Craniological differentiation amongst wild-living cats in Britain and southern Africa: natural variation or the effects of hybridisation? *Animal Conservation,* 7: 339-351.

Yamane A., Doi T. & Ono Y. 1996. Mating behaviors, courtship rank and mating success of male feral cat. *Journal of Ethology*, 14: 35-44.

INDEX

A

B

F

G

H

I

P

Q

R

S

T

U

V

W

Z